ENGINEERING & COMPUTER GRAPHICS WORKBOOK

Using SOLIDWORKS 2019

Ronald E. Barr, Ph.D.
Professor

Thomas J. Krueger, Ph.D.
Senior Lecturer

Davor Juricic, D.Sc.
Professor Emeritus

Alejandro Reyes
MSME, CSWE, CSWI

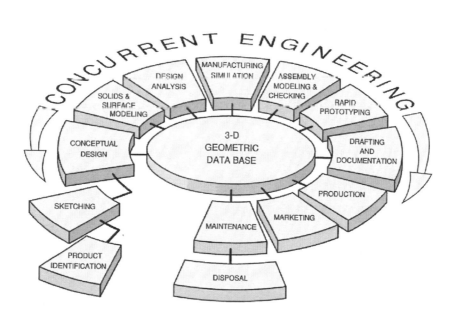

SDC Publications

P.O. Box 1334

Mission, KS 66222

913-262-2664

www.SDCpublications.com

Publisher: Stephen Schroff

ISBN-13: 978-1-63057-219-8

ISBN-10: 1-63057-219-5

Printed and bound in the United States of America.

Table of Contents

1. Computer Graphics Lab 1: 2-D Computer Sketching I

Introduction to SOLIDWORKS; Screen Layout; Main Pull-Down Menu; Feature Manager Tree; View Orientation; View and Display Toolbars; Sketching Toolbars; Sketching Planes; Line Colors; Starting a New Part; Setting Grids and Units; Using Basic 2-D Primitives; Applying Basic Dimensions; Extruding and Revolving Simple Parts; Printing a Hardcopy.

2. Computer Graphics Lab 2: 2-D Computer Sketching II

Review of All 2-D Sketch Entities; Advanced Sketching Tools; Sketch Editing Tools; Linear and Circular Repeats; Basic Dimensioning; Extruding and Revolving Simple Parts.

3. Computer Graphics Lab 3: 3-D Solid Modeling of Parts I

Adding Sketch Relations; 3-D Features Toolbar; Advanced Extrusion and Revolution Operations; Insert Reference Geometry; Mirror 3-D Feature; Create Linear and Circular 3-D Patterns; Building 3-D Solid Parts.

4. Computer Graphics Lab 4: 3-D Solid Modeling of Parts II

Creating Advanced 3-D Features: Draft, Shell, Dome, Loft, Sweep; Advanced Extrusion and Revolution Operations; Building 3-D Solid Parts.

5. Computer Graphics Lab 5: Assembly Modeling and Mating

Building Multiple 3-D Parts; Color Shading of Parts in an Assembly; Starting a New Assembly File; Tiling the Screen Windows; Assembly Toolbar; Drag and Drop Parts into Assembly; Move and Rotate Component; Mate Parts with Different Mate Types; Print Assembly File.

6. Computer Graphics Lab 6: Analysis and Design Modification I

Measure Function; Mass Properties Function; Types of Mass Properties and Applicable Units; Print Mass Properties Report; Design Modification of a Solid Model; Setting Up a Design Table; Setting Parameters for the Design Table; Configuration Manager; Print Assembly File.

7. Computer Graphics Lab 7: Analysis and Design Modification II

Introduction to Finite Element Analysis Using SOLIDWORKS Simulation; Definition of FEA Terms; Building a Solid Model for an FEA Study; Beginning an FEA Study; Applying Loads and Constraints; Creating a Mesh; Analyzing the Model for Stress Distribution; Printing the von Mises Stress Distribution; Design Modification of a Solid Model Based on Analysis Results.

8. Computer Graphics Lab 8: Kinematics Animation and Rapid Prototyping

Introduction to the SOLIDWORKS Animation Wizard; Loading an Assembly File; Exploding an Assembly; Creating the Animation; Animation Controller; Editing the Animation; Saving an .AVI File; Introduction to Physical Simulation, Introduction to Rapid Prototyping; Saving an .STL File; Sample Solid Models for Rapid Prototyping.

9. Computer Graphics Lab 9: Section Views in 3D and 2D

Viewing 3D Section Views of a Solid Model; Printing 3D Section View; Inserting a Drawing Sheet; Setting Drawing and Hatch Pattern Options; Projecting Three Orthographic Views Onto a Drawing Sheet; Creating the Cutting Plane Line; Making a 2D Section View; Completing a Section View Drawing; Print Section View Drawing.

10. Computer Graphics Lab 10: Generating and Dimensioning Three-View Drawings

Inserting a Drawing Sheet; Setting Drawing Sheet Options; Projecting Three Orthographic Views of a Solid Model Onto a Drawing Sheet; Adding Centerlines and Completing the Drawing Views; Setting the Dimensioning Variables; Dimensioning the Drawing; Adding Title Block and Annotations; Print a Drawing.

APPENDIX A – Example of a Titleblock with Dimensions

NOTES:

Computer Graphics Lab 1: 2-D Computer Sketching I

INTRODUCTION TO SOLIDWORKS

SOLIDWORKS© is a parametric solid modeling package that is used to build solid computer models. The designer starts with a 2-D sketch that at first is loosely defined. Dimensions and other constraints are applied to the 2-D sketch to fully define its geometry. The sketch is turned into a 3-D solid model using the extrusion or revolution command. If the base part needs to be altered, the designer can simply change the dimensions of the original sketch or definition of the base feature, and then rebuild the part with a click of the mouse. Once the base part is defined, new sketches can be defined on planes around the base part, and they can then be extruded to form bosses and cuts on the part. Also, special design features can be easily added to the part, such as fillets, chamfers, and counterbores. This is called feature-based modeling and it is an approach to 3-D computer modeling. In this first computer graphics lab, you will concentrate on 2-D computer sketching methods, and will extrude or revolve simple parts. Four different, short exercises are provided.

SOLIDWORKS TOOLBARS SETUP

Your instructor will show you how to launch SOLIDWORKS on your computer. When you first launch SOLIDWORKS the screen will appear as shown in **Figure 1-1**. Your first task is to set up your SOLIDWORKS screen layout to have a common appearance that will be referenced throughout this workbook. To accomplish this, follow these instructions. Just to the right of the SOLIDWORKS icon at the upper left of the screen, there is an arrow. Click on this arrow to get the basic tab menu. To keep this on the screen, **Select** the **stickpin** at the right end of the menu that appeared. This will always keep this menu visible. Go to the **View** pull-down menu on the top of screen, select **Toolbars** and deactivate all the toolbars except **Dimensions/Relations, Features, Sketch, Standard, Standard Views,** and **View**, as shown in **Figure 1-2**. This will arrange the common toolbars around your screen for easy access during your exercises. Note: these toolbars will not appear until you have begun to model a new part. As you get familiar with the screen and the commands, you may choose to customize your toolbars to fit your needs.

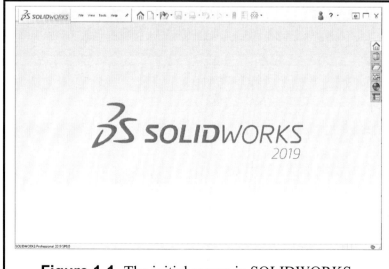

Figure 1-1. The initial screen in SOLIDWORKS.

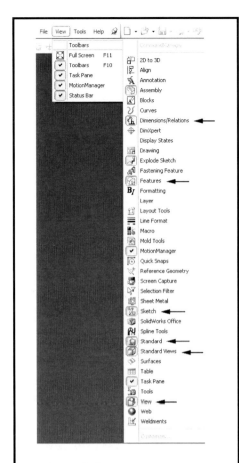

Figure 1-2. Setting the Proper Screen Toolbars during Setup.

THE SOLIDWORKS SCREEN LAYOUT

The SOLIDWORKS toolbars and menus are automatically arranged around the screen layout. A main toolbar called the CommandManager groups many toolbars in the same location using tabs to save screen space. The center of the screen is the sketching area for your design work. Study **Figure 1-4** to become familiar with this screen layout. Each of these menus and toolbars will be described in the subsequent paragraphs.

STARTING A NEW PART

To start your session with SOLIDWORKS, you need to pull down the **File** main menu and click on the **New** command. This will display the New Document dialog, shown in **Figure 1-3**. You have three options: Part, Assembly, and Drawing. To start a new model, click **Part** and then **OK**. Once you have clicked the **OK** button, the SOLIDWORKS interface will be presented with all the previously checked toolbars activated. Your first step is to study the initial screen layout shown in **Figure 1-4**. Some of the toolbars may appear at other positions on the screen.

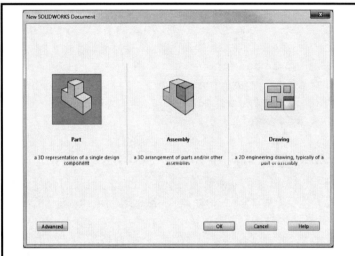

Figure 1-3. Starting a New Part in SOLIDWORKS.

Figure 1-4. The SOLIDWORKS Screen Layout with Major Menus and Toolbars Shown.

MAIN MENU – to keep this menu on the screen, select the stick pen at the right end of the menu bar.

File is where you can create New files and Open, Save, Print Preview, and Print existing files. You also can set up your Page on this menu.

Edit is where you can Undo the last command and Cut, Copy, and Paste Entities. It also has the Rebuild command to update the model when you make changes.

View is where you Redraw and set Display parameters. You can set the View Orientation, and select View commands like Zoom, Rotate, and Pan.

Insert is where the major feature creation commands are located. Here you can create the Base part, Cut the part, and invoke many of the parametric features.

Tools has an assortment of helpful commands. It has Sketching Tools, Dimensioning Tools, and Relations Tools. It also has analysis tools like Mass Properties.

Window is where you can set some window properties like vertical and horizontal tiling.

Help is where you can access on screen listings of the command functions and other information that will help you navigate and learn SOLIDWORKS.

FEATUREMANAGER TREE

The SOLIDWORKS "FeatureManager Tree" is the pane on the left side of the screen used to keep track of the solid modeling process and operations. It lists every feature and their sketch in a chronological tree diagram, where you can edit every feature if needed. The "FeatureManager" is part of every solid model created in SOLIDWORKS. **Figure 1-5** shows an example of the FeatureManager.

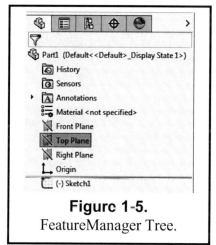

Figure 1-5.
FeatureManager Tree.

THE REBUILD BUTTON

The "**Rebuild Button**" is located on the Standard Toolbar and looks like a red/green traffic light, as shown in **Figure 1-6**. When you select it, the model is rebuilt and updates any changes that have been made. When a model is rebuilt, the current command is terminated.

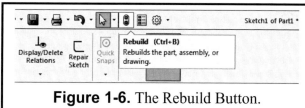

Figure 1-6. The Rebuild Button.

TOOLBARS

Toolbars include the more common commands used in SOLIDWORKS grouped by functionality. To see all of the Toolbars available to you, move your cursor to the blank gray area above the sketch area and click the Right Mouse Button. You can see the toolbars, grouped by function, in the general screen layout of **Figure 1-4** on page **1-2**.

View and **Standard View Toolbars** are shown in **Figure 1-7**. The View Toolbar contains buttons for several Zoom commands, the Rotate View command, and the Pan (Move) View command. It also has options to display your model as a wireframe, with hidden lines, with hidden lines removed, as a shaded image, and as a shaded image with highlighted edges. The Standard View Toolbar allows you to select Front, Back, Left, Right, Top, Bottom, Isometric, and Normal to Plane Views.

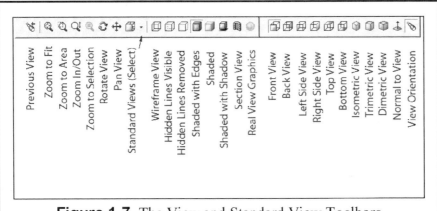

Figure 1-7. The View and Standard View Toolbars at the Top of the Screen.

SKETCH TOOLBAR: The sketch toolbar, shown in **Figure 1-8,** includes the tools to create sketch geometry. The Dimensions and Relations toolbar has the tools to add relations and dimension sketch entities. To start, you click on the **Sketch** icon. As you create 2-D sketch entities, you can use **Dimension** and

Add Relation buttons to define the geometric parameters of your design. The sketch tools available include numerous 2-D commands including **Line, Arc, Circle, and Rectangle**. To edit your sketch, several common commands like **Trim, Mirror, Fillet**, and **Offset** are available.

Figure 1-8. Sketch Toolbar Options.

Select		Centerpoint Arc	
Grid/Snap		Tangent Arc	
Sketch or Exit Sketch		3 Point Arc	
3D Sketch		Ellipse	
3D Sketch On Plane		Partial Ellipse	
Rapid Sketch		Parabola	
Line		Spline	
Corner Rectangle		Spline on Surface	
Center Rectangle		Equation Driven Curve	
3 Point Corner Rectangle		Point	
3 Point Center Rectangle		Centerline	
Parallelogram		Construction Geometry	
Straight Slot		Text	
Centerpoint Straight Slot		Plane	
3 Point Arc Slot		Sketch Fillet	
Centerpoint Arc Slot		Sketch Chamfer	
Polygon		Offset Entities	
Circle		Convert Entities	
Perimeter Circle		Intersection Curve	

SKETCHING PLANES

Before you start a 2-D sketch in SOLIDWORKS, you must select a sketch plane. The three default orthogonal sketch planes are the three standard orthographic view planes: **Front, Top**, and **Right**, as shown in **Figure 1-9**. After a part is created, you can also use any plane or flat surface of the model. The surface does not have to be parallel to one of these three principal planes.

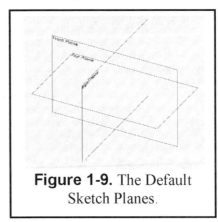

Figure 1-9. The Default Sketch Planes.

SKETCH LINE COLORS

SOLIDWORKS has several ways to give feedback to the user. One of these ways is to use different colors for sketch elements. When sketching 2-D profiles, the following colors are used.

Cyan lines mean that the entities are **Selected**.
Blue lines mean that the dimensions of the geometry are **Under Defined** (undesirable).
Black lines mean that the dimensions of the geometry are **Fully Defined** (preferred).
Yellow lines mean that the dimensions of the geometry arc **Over Defined** (undesirable).
Gray lines mean that these entities are not in the active sketch.

STANDARD TEMPLATES

Before you start any exercises in SOLIDWORKS, it is a good idea to establish some standard 3D model and 2D drawing with title block templates. One set will use the American National Standards Institute (ANSI) in Inches and the other for Metric measurements.

To start, go to **File,** select **New,** then **Part** and **OK** this selection. Go to the **Tools** pull-down menu and select **Options**. In the **Document Properties** tab, make sure that the **Dimensioning Standard** is set for **ANSI** (see **Figure 1-10**). Next, select **Units** in the left-hand column of the menu. In the window that appears, activate **IPS (Inch, Pound, Second)**. You should also set the Decimal Placcs to four **(4)** as indicated in **Figure 1-11**. Close the "Document Properties" menu with the **OK** button. Now go to **File** and select **Save As**. First select **Save as Type** and select **Part Templates (*.prtdot)**, select the **FOLDER** you wish to save this template in, then name the template as **ANSI-INCHES. Make sure that you are saving this into your personal folder**.

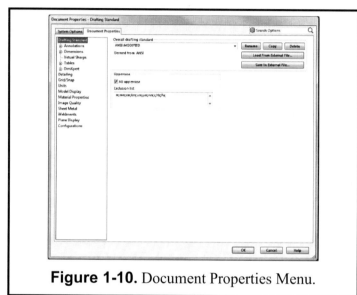

Figure 1-10. Document Properties Menu.

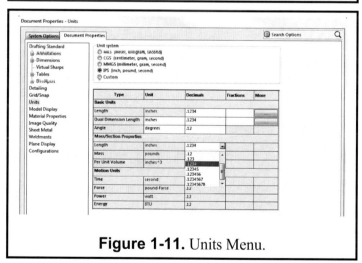

Figure 1-11. Units Menu.

Repeat the above process, only this time, the **Units** you select will be **MMGS (Millimeter, Gram, Second)**. Two decimals should be sufficient for metric measurements. Save this template to your **FOLDER**, change the **Save as Type** to **Part Templates (*.prtdot)**, set your title to **ANSI-METRIC. Make sure that you are saving this into your folder.**

STANDARD TITLE BLOCK DRAWING SHEETS

The next process is to set up a 2D Drawing with Title Block to reflect the two dimensioning standards. You may have to draw a title block (see **Figure 1-12**) if one has not been provided. An example of a title block, with suggested dimensions and style, is available in **Appendix A**.

a) If a 2D Drawing template has not been provided, select the menu **File** and **New**, select the **Drawing** and click OK. Select the size **A-(ANSI) Landscape.** Turn off the **Display Sheet Format** option and click OK. If needed, click on the red **X** to cancel the **Model View** command.

b) If a 2D Drawing template with Title Block has been provided, go to **File** and **Open** that drawing file.

Go to the **Tools** pull-down menu and select **Options**. Under the **Document Properties** tab make sure that the **Dimensioning Standard** is set for **ANSI**. Next, select **Units** in the left-hand column. In the window that appears, activate **IPS (Inch, Pound, Second)**. Three **(3)** decimal places will be all you should use here unless you are doing very high tolerance drawings.

To draw the Title Block, right click on **Sheet1** in the **FeatureManager** and select **Edit Sheet Format**. Use the Sketch and Annotations toolbar tools to draw the title block. Go to **Insert – Annotations - Note** to add your personal information, such as **NAME, DESK NUMBER**, and **SECTION NUMBER**. When finished, in the **FeatureManager Tree**, right click on **Sheet1** and select **Edit Sheet**. Now that this is done, go to **File** and **Save As**. First select the **Save as Type** and select **Drawing Template (*.drwdot)** and select the folder you wish to save this template in, then title your part as **TITLEBLOCK-INCHES. Make sure that you are saving this into your personal folder.**

Repeat the above process, only this time, the **Units** you select will be **MMGS (Millimeter, Gram, Second)**. Two **(2)** decimals should be sufficient for metric measurements. Add your personal information such as **NAME, DESK NUMBER**, and **SECTION NUMBER**. In the **FeatureManager Tree**, right click on **Sheet 1** and select **Edit Sheet**. Save this template in the same way, but change your title to **TITLEBLOCK-METRIC** and select the **Save as Type** and select **Drawing Template (*.drwdot). Make sure that you are saving this into your personal folder.**

INSERTING A PART (3D MODEL) ONTO A TITLE BLOCK (2D DRAWING) SHEET

For each of the assignments you will be asked to submit a printed copy of the part that you built in SOLIDWORKS. The following will be the procedure to accomplish this requirement.

A. Build the solid model according to instructions and save that part in your folder with the file extension of **.sldprt**.

B. Open the appropriate Title Block drawing sheet (see **Figure 1-12** for example). The Title Block drawing sheet that you open should have the same units as the solid part (metric or inches).

C. Pull down **Windows** and then select **Tile Vertically**. This allows you to see all the windows that you have open at the same time.

D. Select the 3D Part Name in the **FeatureManager Tree** and while holding down the left mouse button, drag it onto the Title Block drawing sheet. At this time you have several options available to choose from. For most of your labs, in the **Orientation** options turn off **Create Multiple Views** and select **Current Model View**. In unit nine and ten you will be choosing the three views layout of the 3D model that you insert onto the Title Block. If you do not see the model view, it means you are editing the **Sheet Format**. Right click on **Sheet1** in the **FeatureManager Tree** and select **Edit Sheet**.

E. Select the window of the 3D model view. In the drawing view's PropertyManager pane on the left of the screen, you can adjust the scale of the image to better fill the sheet.

F. In the drawing area of the Title Block Sheet, **Insert** the title of the object. Provide the exercise number in the upper right-hand box of the Title Block. Your Title should be placed in an open area of the Title Block and a larger font size, i.e., **20 to 24 PT OR .25"**. Your name, desk, Sec and exercise number should be font size **12pt or .125"**.

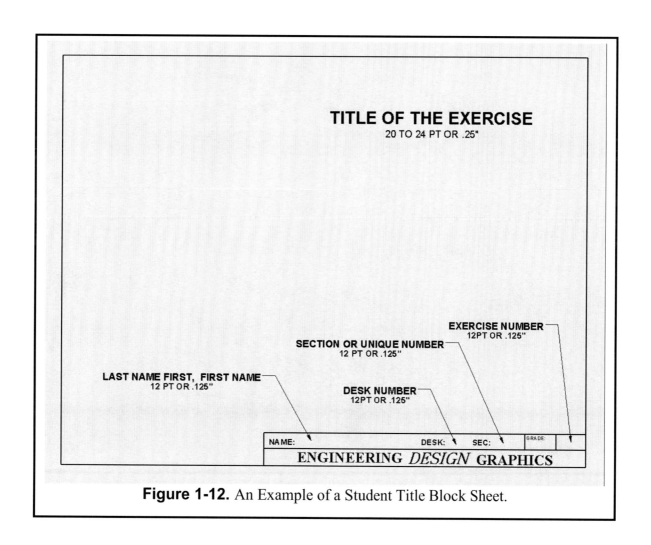

Figure 1-12. An Example of a Student Title Block Sheet.

NOTES:

Exercise 1.1: CARBON FIBER GASKET

In this first exercise, you will build a Carbon Fiber Gasket part. This part is primarily a 2-D sketch with no thickness that will be extruded in 3D. The gasket's sketch will be drawn using many of the 2-D sketching tools and editing commands available in the sketch toolbars. Go to your folder and **Open** the file **ANSI-INCHES.prtdot.** In order to avoid corrupting the **ANSI-INCHES.prtdot** template go to the menu **File - Save as:** under **File Name**, type **CARBON FIBER GASKET**, and under **Save as type**, select **.sldprt** and select **SAVE.**

First set up the sketching grid and units. In the **Tools** menu select **Options.** Under the **Document Properties** tab, click the **Grid/Snap** tab and click **on** (√) all the "Grid" boxes. Set the "Major grid spacing" to **1.00** and "Minor lines per major" to **4**, and set "Snap points per minor" to **1**, as shown in **Figure 1-13**. Click on the **Go To System Snaps** and make sure that the **Grid** and **Snap only when grid is displayed** boxes are checked. Also make sure the **Units** are in **Inches** to **3** decimal places. Then click **OK** to close the menu.

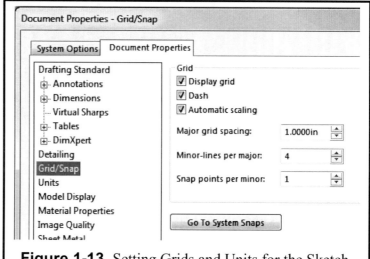

Figure 1-13. Setting Grids and Units for the Sketch.

You will design the gasket in the front view, so click on the **Front Plane** in the FeatureManager tree. The front plane should highlight in cyan. Next, activate the **Sketch Tab** in the CommandManager and select the **Sketch** icon just above and to the left of the Sketch Tab. The sketching grid will now appear on the screen with grids spaced every 0.25 inches. Also make sure you are viewing this from the Front View orientation (see **Figure 1-7**).

Now study the outline of the gasket shown in **Figure 1-14**. You will sketch this outline on the quarter-inch grids using the line tool. Click the **Line** icon on the sketch menu toolbar. Now draw five lines according to the specifications in **Figure 1-14** and detailed below. All **(X,Y)** coordinates are relative to the sketch "origin" seen in the middle of the screen. You are able to track your coordinates at the bottom-right of the sketch screen.

> The *top line* starts at **(-3.00, 1.00)** and ends at **(3.00, 1.00)**.
> The *right-side line* starts at **(3.00, 1.00)** and ends at **(3.00, -1.50)**.
> The *bottom right diagonal line* starts at **(3.00, -1.50**) and ends at **(0.00, -3.50)**.
> The *bottom left diagonal line* starts at **(0.00, -3.50)** and ends at **(-3.00, -1.50)**.
> The *left side line* starts at **(-3.00, -1.50)** and ends at **(-3.00, 1.00)**.

The 2-D outline of the gasket should now be finished and completely enclosed.

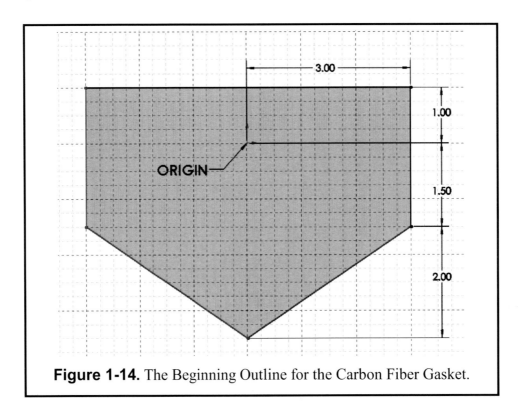

Figure 1-14. The Beginning Outline for the Carbon Fiber Gasket.

The five corners of the gasket are sharp and need to be rounded. This can be easily accomplished with the **Sketch Fillet** command. Click the **Sketch Fillet** icon (round corner symbol) on the Sketch toolbar. The "Sketch Fillet" PropertyManager appears where the FeatureManager tree usually is displayed (**Figure 1-15**). Enter a fillet radius of **0.50** inches. Now, one by one, click on the two edges of the corners (or the endpoint at the intersection) to create a fillet there. Once the five fillets are added, click on the green (√) OK button to continue. The five fillets are now completed as shown in **Figure 1-16**.

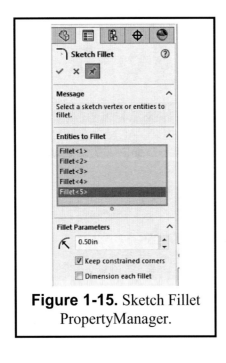

Figure 1-15. Sketch Fillet PropertyManager.

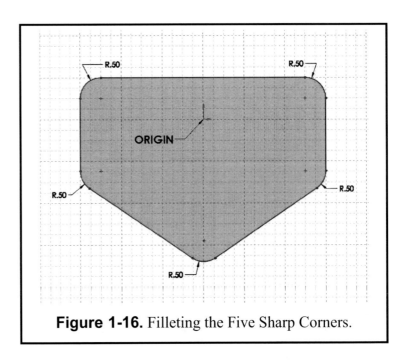

Figure 1-16. Filleting the Five Sharp Corners.

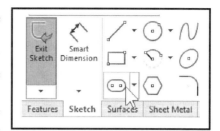

Using the **Straight Slot** tool, draw a slot. Locate the first center at **(-1.5, -1.5)** then drag your cursor to **(1.5, -1.5)**. After you select the second point, move your cursor vertically to form the slot. If the slot is automatically dimensioned delete the slot dimensions. **Select** the **Smart Dimension** tool and dimension the arc of the slot to be **0.25"** radius to complete the slot. Next, select the **Circle** sketching tool, and create the five circles concentric to the fillets (rounded corners) of the outline. All circles have the same diameter of **0.375** inches. *Note:* After drawing the five circles use the Smart Dimension to dimension the circles to make sure that they have a diameter of **0.375** inches. **Note:** While you are sketching the lines, rectangle, or circles, the end of the cursor shows a small icon that indicates the current type of sketch entity that you are using. This is called a "smart cursor." A small grid icon shows when the grid snap is activated.

Many models have symmetrical features and with the following commands those features become very easy to replicate. You will now draw two polygons on the gasket that are symmetrical about a centerline. First select the **Centerline** command under the **Line** menu and draw a vertical centerline through the **Origin**. Next, pull down **Tools**, select **Sketch Tools**, and then select **Dynamic Mirror.** The centerline will now have an equal sign at either end to indicate the Dynamic Mirror is active. Next, select the "**Polygon**" command icon with a hexagon in the sketch toolbar. The Polygon's PropertyManager options are displayed. Draw the polygon on the gasket to the right side of the centerline. First, **locate the center** of the Hexagon. Then **move your cursor horizontally** from that center to locate a vertex of the polygon. Now study the actual parameters needed for the polygon and then set them as shown in **Figure 1-17**.

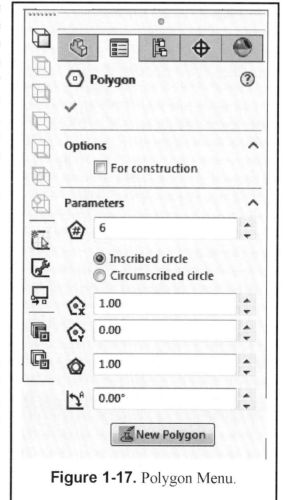

> Number of sides = **6**
> Select - Inscribed circle
> X-origin = **1.00**
> Y-origin = **0.00**
> Inscribed circle diameter = **1.00**
> Rotation angle = **0°**

Once the parameters are set, click on the green (√) OK button in the "Polygon" PropertyManager to finish the polygons, as shown in **Figure 1-18**. This completes the sketch, and now can it can be extruded to create the solid model.

Figure 1-17. Polygon Menu.

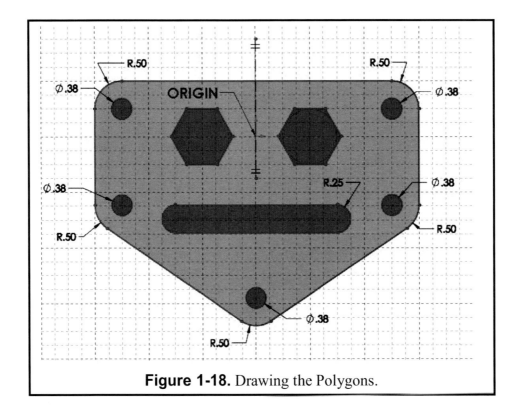

Figure 1-18. Drawing the Polygons.

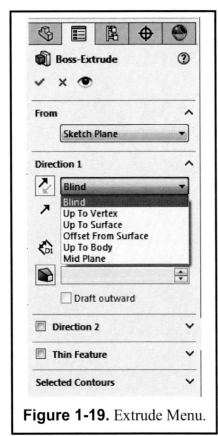

Figure 1-19. Extrude Menu.

EXTRUDING A TWO DIMENSIONAL SKETCH

The **Extruded Boss/Base** command is found in the **Features** tab. This command will extrude a sketch or selected sketch contours in one or two directions to create a solid feature. Under **Direction 1**, **Blind** is the default selection; however, the drop-down list reveals additional choices as shown in **Figure 1-19**. The extrusion options you will be using are **Blind**, **Up to Surface** and **Mid Plane**. The **Blind** option will extrude the sketch a specified distance. To change the direction of the extrusion, click on the Reverse Direction button next to the "**Blind**" box (See **Figure 1-20**). **Up to Surface** will extrude the sketch up to the selected surface. **Mid Plane** will extrude the sketch equally in both directions away from the sketch plane.

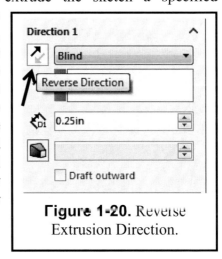

Figure 1-20. Reverse Extrusion Direction.

EXTRUDE THE GASKET SKETCH

Now extrude the gasket to its needed thickness. Select the **Features** tab in the CommandManager and select **Extruded Boss/Base** or pull down the **Insert** main menu, select **Boss/Base**, and then **Extrude**, as shown in **Figure 1-19**. The "Extrude" properties are displayed in the PropertyManager area (**Figure 1-21**). Set the parameters for the extrusion:

> Direction One = **Blind**
> Distance1 = **0.25** inches

The view is rotated to an Isometric orientation, and a preview of the extrusion is shown in the screen. Click on (√) the green button on the "Extrude" properties to finish.

Now you can view your model. On the "View Orientation" window click **Trimetric**. You should see the solid model in a trimetric orientation (see **Figure 1-22**). Also try **Isometric**, **Dimetric**, and **Front** view orientations. You can experiment with some of the viewing operations in SOLIDWORKS. Click the **Rotate View** icon (it appears as circular arrows on the top toolbars) or press down on the center wheel of the mouse and drag the model in the screen. Now rotate your model into different configurations. Also try several **Zoom** commands and the **Pan** command.

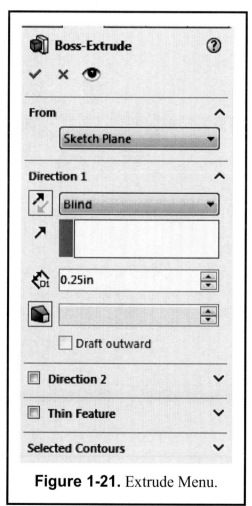

If you would like to change the color of your model, right click on the model name in the FeatureManager tree and then select the Appearances icon (colored ball) in the context menu. You can then assign any color you wish to the model.

You should save your model. Pull down the **File** menu and select **Save As**. On the "Save As" menu, select your appropriate file folder in your account, type in the part name **CARBON FIBER GASKET** and select the file type as **Part (*.sldprt)**, then click **Save**.

Before you finish this exercise, you should print a hard copy for submission to your instructor. *Check with your instructor first for any special printing instructions.* You may need to use a special sheet like the Title Sheet shown in **Figure 1-12**. If you open **TITLEBLOCK-INCHES.DRWDOT** you must do a **Save As** and save the new Drawing with the name that you want to use (**CARBON FIBER GASKET**) and select the file type as **Drawing (*.SLDDRW)**. You will now be able to insert the **CARBON FIBER** 3D model view into your **Title Block** drawing sheet that was created in the introduction to this chapter, as shown in **Figure 1-22**. To put this object on a Title Sheet, follow the instructions on **Page 1-7**.

Figure 1-21. Extrude Menu.

If you check (√) **off** the "Use documents Font" in the properties, you can change the font, for example, to:

> Font = **Arial**
> Font Style = **Regular**
> Points = **12**

Place the title of the part in a convenient location of the title sheet. The title should be at 24 pts or .25" in height. Also place the exercise number in the blank to the right of the GRADE box.

Now **SAVE** your drawing sheet to your directory as **CARBON FIBER GASKET.slddrw**. Take note that the title of the part and the drawing are the same, but the extension is different. The 3D solid model and the 2D drawing are linked, meaning that any changes made to the solid model will automatically be updated on the drawing.

Before you print your copy, make sure you have set the **File**, **Page Setup** to **Landscape** mode. Now pull down the **File** and **Print Preview** commands from the top main menu, then the **Print** button to send it to the default printer.

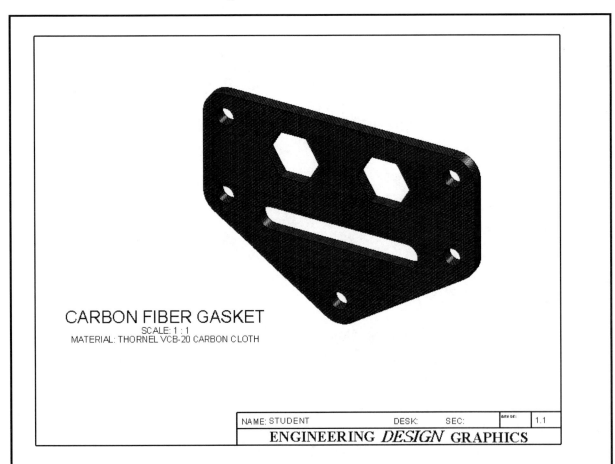

CARBON FIBER GASKET
SCALE: 1 : 1
MATERIAL: THORNEL VCB-20 CARBON CLOTH

| NAME: STUDENT | | DESK: | SEC: | GRADE: | 1.1 |

ENGINEERING *DESIGN* GRAPHICS

Figure 1-22. CARBON FIBER GASKET Model view added to the Title Block Drawing Sheet.

Exercise 1.2: COVER PLATE

There are many different approaches to creating a 2-D computer sketch using **SOLIDWORKS**. The approach will depend on the design intent and also on the designer's own preferences and choices. In the previous Exercise 1.1, geometry was constructed in the Front plane using X-Y coordinate geometry and a grid system. In this Exercise 1.2, you will construct the 2-D geometry on a horizontal (Top) plane since this is the normal orientation of the cover plate. Also, instead of using a grid coordinate system for laying out the design, you will use dimensions to define the geometry about the sketch origin.

Go to your folder and **Open** the file **ANSI-INCHES.prtdot**. In order to avoid corrupting the **ANSI-INCHES.prtdot template** go to **File - Save as:** under **File Name**, type **COVER PLATE**, and under **Save as type**, select **.sldprt** and select **SAVE.** Click the **Top plane** in the

"FeatureManager." Also click **Top** view in the Standard View toolbar to see the top plane in *cyan* on the screen. Now click the **Sketch** tab and **Select** the **Sketch Icon**, then click the **Circle** icon. Draw a large circle centered at the origin. For now, the diameter is unimportant because you will dimension it. Select the **Smart Dimension** command (slanted dimension line) in the Sketch tab. Click to select the circle and click to locate the dimension on to the upper right side of the circle. The dimension is not likely to be correct; as soon as you place the dimension a small Modify box appears (see **Figure 1-23**). Key in **10.00** inches and click the green (√) check box. The diameter of the circle changes to the new value. If the circle gets too big, use **Zoom** or Select the **Top View** again to see it better.

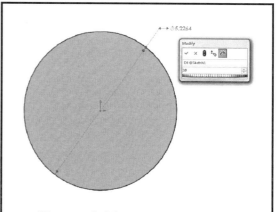

Figure 1-23. Setting the Circle Diameter.

Now you will extrude the sketch circle into the solid base part of the cover plate. Select the **Features** tab and select **Extrude** or pull down **Insert**, select **Boss/Base**, and then **Extrude.** Set the parameters for the extrusion to be:

    Direction One = **Blind**
    Distance1 = **0.50** inches

Click the green (√) button on the "Extrude" command to finish. View the cover plate as an **Isometric (Figure 1-24)**.

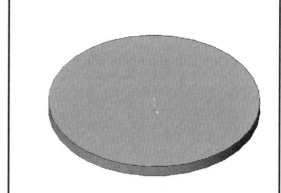

Figure 1-24. Base Part of the Cover Plate.

You will now draw the sketch geometry for the indentation of the cover plate. **Click on the top surface of the Cover Plate model** - it should turn *blue*. This will be the next sketch surface. Also select **Top** on "View Orientation" for better viewing of the sketch plane. Select the **Sketch Tab** and then click on the **Sketch** icon to start the next sketch. Select the **Circle** tool and draw a circle centered at the origin. Now draw a second bigger circle centered at the origin but <u>not</u> as big as the 10.00 inch diameter of the base plate. Now, using the **Smart Dimension** icon, set the diameter of the smaller new circle to **3.50** inches and the diameter of the bigger new circle to **8.50** inches.

Using the **Line** sketch tool, draw a horizontal line from the origin toward the right. Next, draw a line from the origin at an angle above the horizontal line. Both should span across the two circles as shown in **Figure 1-25**. Next, **Dimension** the angle to **22.5 degrees**. Select the horizontal line and then the angled line to add the angular dimension (**Figure 1-25**).

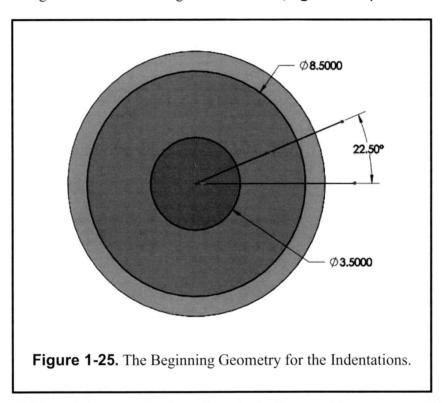

Figure 1-25. The Beginning Geometry for the Indentations.

Now select the **Trim** tool from the sketch toolbar. In the PropertyManager you will see the **Trim command options**. Select the **Trim to Closest** option. A scissors icon now appears on the end of the cursor indicating that trim command is active. Trim the line portions (two places) that extend beyond the big circle by just clicking on them with the **LMB**. Next, trim the two line pieces from the origin to the small circle. Finally, trim the portions of the circles that lie outside of the angle. Your sketch should now look like **Figure 1-26**. Go to the **Linear Sketch Pattern** pull down command and **Select Circular Sketch Pattern**. Activate the first option box and **Select** the **Origin**. Change the number of instances to **8**. The last step is to activate the **Entities to Pattern** selection box and select the four sketch elements **(See Figure 1-27)**. Click OK to finish. The completed sketch for the indentations is shown in **Figure 1-28**.

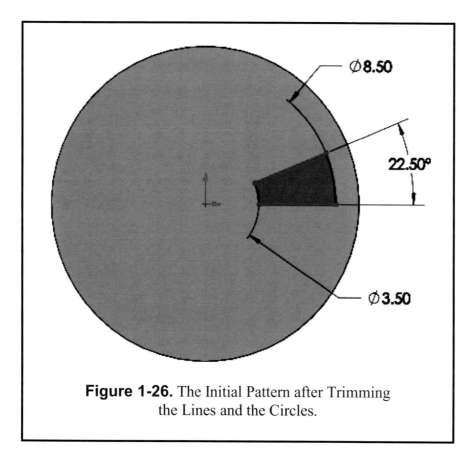

Figure 1-26. The Initial Pattern after Trimming the Lines and the Circles.

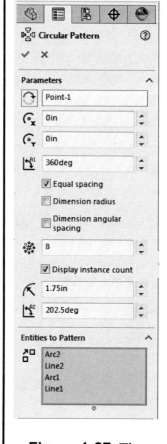

Figure 1-27. The Circular Pattern.

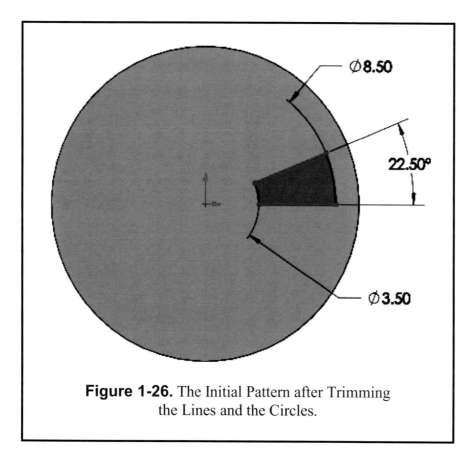

Figure 1-28. The Finished Pattern for the Indentations.

Now you will cut the indentation sketch into the cover plate. It is helpful to change **to an Isometric** view orientation when adding a 3D feature. From the **Features** tab select the **Extruded Cut** command. In the Cut-Extrude PropertyManager enter the following parameters:

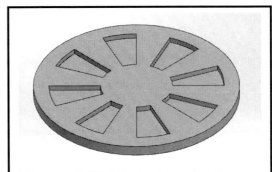

Direction One = **Blind**
Distance1 = **0.25** inches

Click the green (√) **OK** button to finish. The finished part is shown in **Isometric** (**Figure 1-29**).

Figure 1-29. The Indentation Pattern Cut into the Plate.

We will now create the lift hole in the middle of the plate. Change to a **Top** view orientation and select the top face of the cover plate (it should turn *blue*). You might also need to **Zoom** in to see the center portion better. Click the **Sketch Tab** and then select the **Sketch** icon to get into the sketching mode. Since the lift hole is symmetric about the origin, you can use a mirror function. Start by drawing a vertical **Centerline** (CL) through the origin. Next, go to **TOOLS – SKETCH TOOLS - DYNAMIC MIRROR**. There will be an equal sign on either end of the centerline you drew. Every operation you perform on the right side is duplicated on the left side. Now, draw a small **Circle** to the right side of the centerline. Then draw two parallel **Line**s that start at the centerline and go through part of the circle, as indicated in **Figure 1-30**. Now add the following **Dimensions** to the sketch geometry.

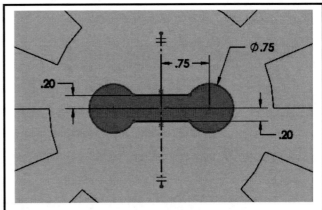

Circle Diameter = **0.75"**
Lateral Dist. from CL = **0.75"**
Line1 from origin = **0.20"** (above)
Line2 from origin = **0.20"** (below)
Circle center = **horizontal with origin**

Now **Trim** the two line pieces that overhang the circle, and then trim the part of the circle between the two parallel lines. If you get a warning from SOLIDWORKS about eliminating associated dimensions while you are trimming, just ignore it and answer **Yes** to proceed with the trim.

Figure 1-30. Geometry and Dimensions for the Lift Hole.

Cut the lift-hole sketch through the cover plate. Change to an **Isometric view** to make the Extruded Cut. From the **Features** tab select **Extruded Cut** or from the menu **Insert**, select **Cut**, and then **Extrude**. In the Cut-Extrude PropertyManager select **Through All** from Direction1. Click the green (√) **OK** button to finish this extrude-cut operation and view the lift hole in **Figure 1-31**.

To finish the cover plate, you need to add four circles around the perimeter to make the guide holes. Select the top surface again and **Sketch** four **Circles** around the perimeter of the cover plate. The circles should be at 90 degrees to each other. The centers should be on the perimeter and aligned with the origin. The diameter of all four circles should be **0.75**, as shown in **Figure 1-31**.

OR

Since you have already used the Circular Sketch Pattern, you could draw a **Circle** with a diameter of **.75** and use the **Circular Sketch Pattern** with the center at the **Origin** and the instances set at **4**.

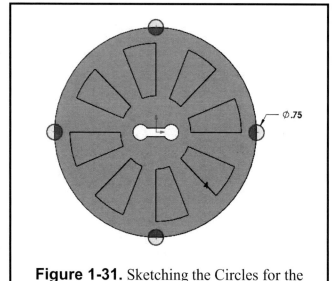

Figure 1-31. Sketching the Circles for the Guide Holes.

Go to the Features tab and select the Extruded Cut command. For Direction1, select the **Through All** end condition, and click the green (√) **OK** button to finish the part. Now view your model in an **Isometric** view, as shown in **Figure 1-32**.

If you would like to change the color of your model, right mouse click in the model name in the FeatureManager tree and select the Appearances command in the context menu. You can then assign any color you wish to the model.

You should now save your model. Pull down **File,** select your file folder, select **Save As,** type in the part name **COVER PLATE.sldprt**, then click **Save**. Insert the isometric view of the Cover Plate into your **Title Block** drawing sheet that was created in the introduction to Chapter 1 (See **Figure 1-32**. Follow the instruction given on **Page 1-7**. Now **SAVE** your drawing sheet to your directory as **COVER PLATE.slddrw**. Take note that the title of the part and the drawing are the same but the extension has changed. The solid model and the drawing are linked, meaning that any changes made to the solid model will automatically be updated on the drawing.

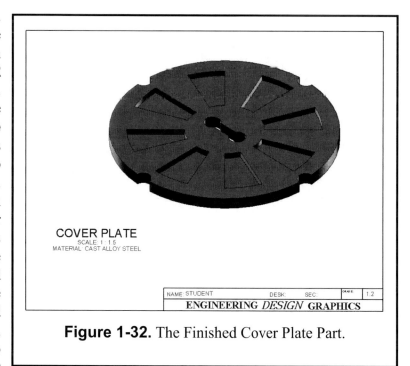

COVER PLATE
SCALE: 1 : 1.5
MATERIAL: CAST ALLOY STEEL

| NAME: STUDENT | DESK: | SEC: | GRADE | 1.2 |
| ENGINEERING *DESIGN* GRAPHICS |

Figure 1-32. The Finished Cover Plate Part.

Exercise 1.3: WALL BRACKET

The sketch geometry used in Exercises 1.1 and 1.2, like line, circle, rectangle, and arc, are sufficient for designing many different kinds of parts. Sometimes more advanced 2D geometry, like an irregular curve, may be needed to complete a design. In the case of the Wall Bracket for this Exercise 1.3, you will learn how to use a spline.

Go to your folder and **Open** the file **ANSI-INCHES.prtdot**. In order to avoid corrupting the **ANSI-INCHES.prtdot** template go to **File - Save as:** under **File Name**, type **WALL BRACKET**, and under **Save as type**, Select **.sldprt** and select **SAVE.**

Select the **Front** plane in the FeatureManager. Also click **Front** view in the Standard View toolbar to see the front plane in *blue* on the screen. Now select the **Sketch** command from the Sketch tab. First sketch the horizontal and vertical lines of the Wall Bracket outline using the **Line** tool. Refer to **Figure 1-33** for the proper **Dimension** for each line segment. Units are in inches and the upper left corner is at the origin.

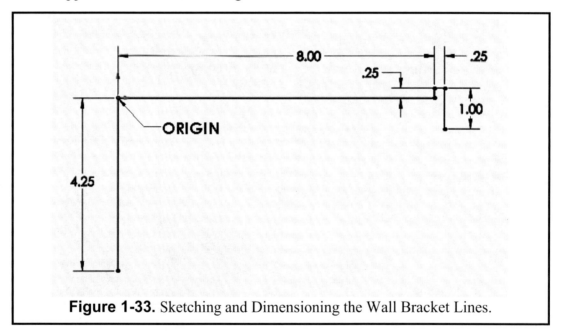

Figure 1-33. Sketching and Dimensioning the Wall Bracket Lines.

Now you will draw a spline from the bottom of the 4.25 inch line to the bottom of the 1.00 inch line. Select the **Spline** command from the Sketch tab (it looks like a sine wave). Click the **LMB** at the first bottom point. Then with the mouse pressed down, drag the spline over to the first intermediate point about one third of the way up and about one third of the way to the right. Release the **LMB** there, then press it down again and drag the spline to the second intermediate point about two thirds of the way up and over. Again release the **LMB** there, then press it down again and drag the spline to the final point, where you end the spline by double clicking the **LMB** or press the ESC key. Refer to **Figure 1-34** for these four points' positions.

The two intermediate points are used to control the shape of the spline. With the **Select** cursor, select the spline (it should highlight *blue* with its four points identified). Now select intermediate point 1 and drag it out and down a little to create a slight bulge. Next, select intermediate point 2 and pull it up a little to create a slight inflection (see **Figure 1-35**). <u>Note</u>: There are *no exact* X-Y coordinates for you to drag these intermediate points to. Just try it a few times until you achieve a shape that pleases you (for example, the shape in **Figure 1-35** is acceptable).

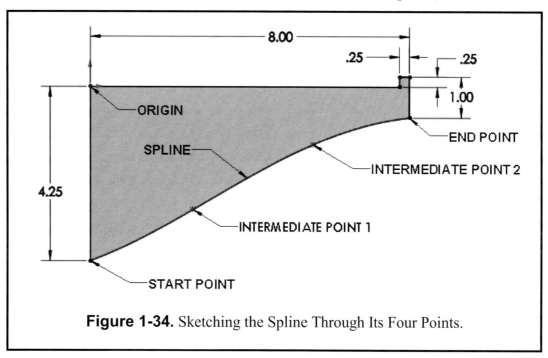

Figure 1-34. Sketching the Spline Through Its Four Points.

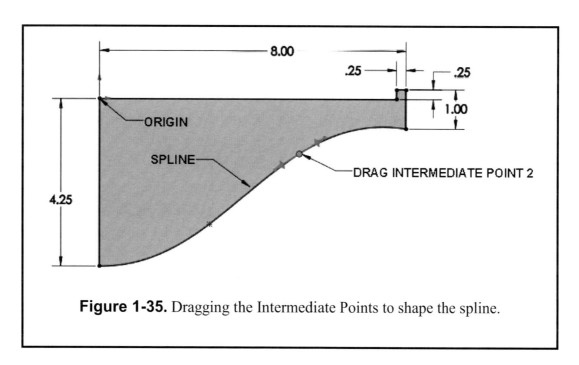

Figure 1-35. Dragging the Intermediate Points to shape the spline.

Now you will complete the left end of the Wall Bracket by adding the wall hooks. On the left edge, add the two wall hooks using the **Line** tool and the **Dimension** values given in **Figure 1-36**. You will have to **Trim** two places on the left edge to make the wall hooks contiguous with the rest of the profile. *Note:* The bottom wall hook has the same dimensions as the top one.

You will finish the Wall Bracket by adding a fillet to the sharp bottom corner on the right end. Since it was earlier associated with the *1.00 inch* dimension, **SOLIDWORKS** will give you a warning if you try to fillet it. So it is best to just first **Delete** that dimension. Select the **Fillet** command. Enter a radius of **0.25** inches (refer back to **Figure 1-16**). Next, select the vertical 1.00 inch right edge of the bracket and pick the spline close to its end point, and then click OK to finish the fillet command. The 2D sketch is now complete and ready to be extruded. Select the **Features** Tab and **Select - Extruded Boss/Base**. Enter a **Blind** distance of **0.125** and click OK to complete the Wall Bracket, as shown in **Figure 1-37** in a **Trimetric** view.

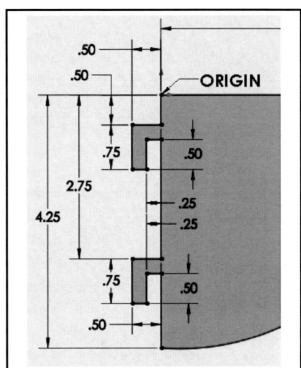

Figure 1-36. The Dimensions Used to Create the Wall Hooks on the Left Side.

If you would like to change the color of your model, right click on the model name in the FeatureManager tree and then select the Appearances colored ball in the context menu. You can then assign any color you wish to the model.

Now **Save** the part as **WALL BRACKET.sldprt**. Insert the Wall Bracket onto a Title Block drawing sheet. Save as **WALL BRACKET.slddrw** and **Print** a hard copy for your instructor. See Page 1-7 for instructions.

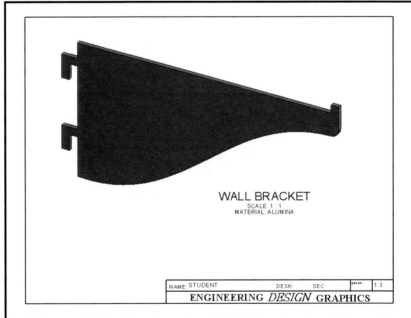

WALL BRACKET
SCALE: 1 : 1
MATERIAL: ALUMINA

| NAME: STUDENT | DESK: | SEC: | | 1.3 |
| ENGINEERING *DESIGN* GRAPHICS | | | | |

Figure 1-37. The Completed Wall Bracket in a Trimetric View.

Exercise 1.4: MACHINE HANDLE

In the previous exercises, the parts were designed using English units (inches). International System (SI), or Metric units are also common in engineering practice. In Exercise 1.4, the Machine Handle will be designed using millimeters as the basic units (see **Figure 1-38**). Since the machine handle is a cylindrical body, you will create a **Base Revolve** rotating a 2-D sketch about an axis of revolution.

Go to your folder and **Open** the file **ANSI-METRIC.prtdot**. In order to avoid corrupting the **ANSI- METRIC.prtdot** template go to **File - Save as:** under **File Name**, type **MACHINE HANDLE**, and under **Save as type**, Select **.sldprt** and select **SAVE.**

Next, Go to **TOOLS - OPTIONS – DOCUMENT PROPERTIES** and Select "**Grid/Snap.**" Make the following settings on this menu and click OK to finish:

"Major grid spacing" = **20 mm**
"Minor lines per major" = **4**

Also go to **System Snaps** and make sure "**Grid**" and the "**Snap**" functions are checked (√) on, then click the **OK** button.

In the **FeatureManager Tree**, select the **Front** plane and from the **Sketch Tab** select the **Sketch** command to start your sketch. The screen area should show a metric grid with spacing every 5 millimeters. You may need to **Zoom** in. Using the **Centerline, Line**, and **Centerpoint Arc** tools, draw the closed 2-D sketch profile shown in **Figure 1-38**.

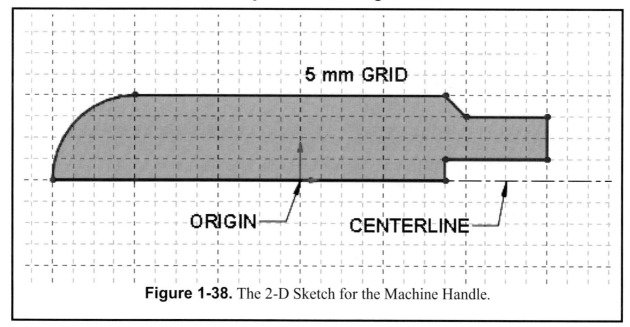

Figure 1-38. The 2-D Sketch for the Machine Handle.

Now you will revolve this sketch into a solid model. Select the **Features** tab and click in the **Revolved Base/Boss** command. The "Base Revolve" PropertyManager is presented as shown in **Figure 1-39**. The main choice you have here is the degrees of revolution about the centerline. Set it to a full revolution of **360°** and then click OK to finish. Your part should now look as shown in **Figure 1-40** with an **Isometric** view.

There are many different designs for a Machine Handle, and the one here is quite typical. After studying it, however, you notice that the sharp outer edge on the right side poses a concern. SOLIDWORKS allows you to easily fix these design changes after the solid model has been built. One way is to simply go back and edit the original sketch.

Notice in the FeatureManager that SOLIDWORKS keeps a chronological list of the features you used to model the part. One of them should read "**Revolve**," which is the operation you just completed. Left of this "**Revolve**" is a right facing arrow. Click on this arrow and down comes the name of "**Sketch1**." Right mouse click on Sketch1 and a context menu appears as shown in **Figure 1-41**. Select the **Edit Sketch** command to continue editing the sketch, with grids, on the screen. Select the **Front** view orientation to better see the sketch.

There are different ways to eliminate the sharp edge. One way is to use the Sketch Chamfer command, which is nested with the fillet icon. A **down arrow** next to the **Sketch Fillet** command shows a pull-down menu and in that menu is the **Sketch Chamfer** command. The "Sketch Chamfer" PropertyManager is displayed. Set the "Chamfer parameters" as shown in **Figure 1-42**, including the chamfer distance of **3 mm**. Now select the two lines that form the sharp edge corner, and click OK to complete the Sketch Chamfer operation.

After the chamfer is added to the sketch, you need to rebuild the model to update the 3D model. Pull down **Edit**, and select the **Rebuild** option (it may be in the *customize menu* section of the pull-down menu).

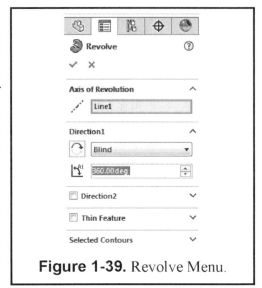

Figure 1-39. Revolve Menu.

Figure 1-40. The Part after the Sketch is Revolved.

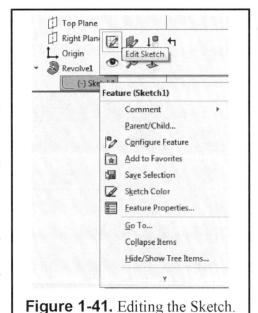

Figure 1-41. Editing the Sketch.

Note: There is also a small icon for the **rebuild command. It appears on the main toolbar and looks like a green and red traffic light**. The model is rebuilt now with a chamfer, as shown in **Isometric** view with the **Shadows in Shaded** mode active **on** (**Figure 1-43**). Now you see how easy it is to modify the sketch of a 3D solid model. You just simply edit the sketch that created the model in the first place.

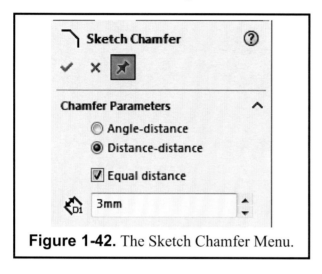

Figure 1-42. The Sketch Chamfer Menu.

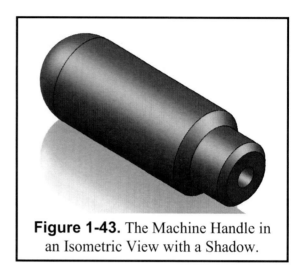

Figure 1-43. The Machine Handle in an Isometric View with a Shadow.

If you would like to change the color of your model, right click on the model name in the FeatureManager tree and select the Appearances command (colored ball) from the context toolbar. You can then assign any color you wish to the model.

You should now save your model. Pull down **File,** select **Save As,** type in the part name **MACHINE HANDLE.sldprt**, and then click **Save**. Insert the isometric view of the Machine Handle model into your **Title Block** drawing sheet, as shown in **Figure 1-44**. Add the title of the part in the open area of the drawing with a 20pt or 6mm letter height. Also add the Exercise number (**1.4**) in the upper right-hand box of the Title Sheet. Now **SAVE** your drawing sheet to your directory as **MACHINE HANDLE.slddrw**.

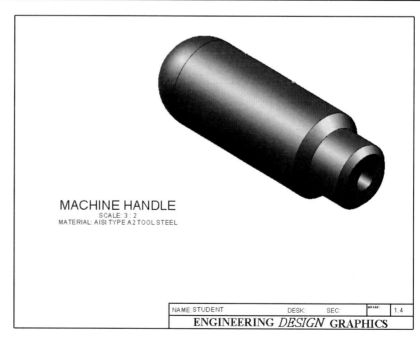

MACHINE HANDLE
SCALE: 3 : 2
MATERIAL: AISI TYPE A2 TOOL STEEL

| NAME: STUDENT | DESK: | SEC: | GRADE: | 1.4 |

ENGINEERING *DESIGN* GRAPHICS

Figure 1-44. A print Preview of the Machine Handle on the Title Block Drawing Sheet.

Supplementary Exercise 1-5: SLOTTED BASE

Make a sketch of the figure below in the **Top Plane** and extrude it **0.50** inches. Add an Isometric view on a Title Block and name it **"SLOTTED BASE."**

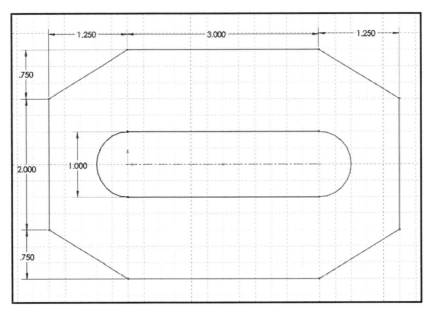

ASSUME THE GRID DIVISIONS TO BE 0.25 INCHES.

Supplementary Exercise 1-6: TRANSITION LINK

Make a sketch of the figure below in the **Front Plane**. Extrude it **0.375"**. Add an Isometric view on a Title Block and name it **"TRANSITION LINK."**

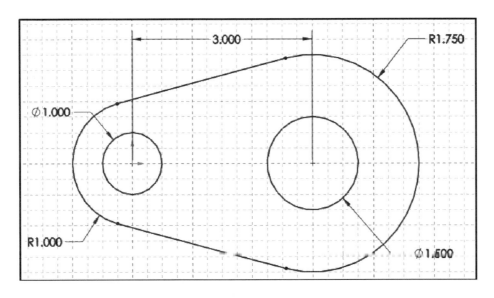

ASSUME THE GRID DIVISIONS TO BE 0.25 INCHES.

Supplementary Exercise 1-7: TEE BRACKET

Make a sketch of the figure below in the **Top Plane** using the ANSI-METRIC template, and extrude it 40 mm as indicated. Add a Trimetric view on a Title Block and name it **"TEE BRACKET."**

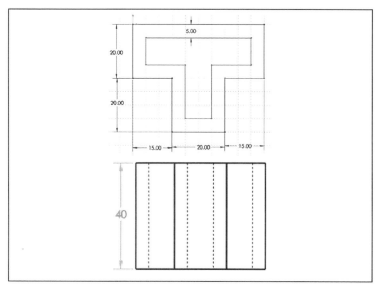

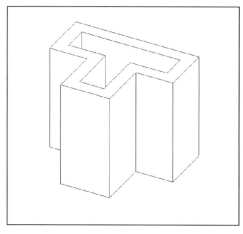

TRIMETRIC VIEW OF THE TEE BRACKET

ASSUME THE GRID DIVISIONS TO BE 5 MM.

Supplementary Exercise 1-8: CABLE SPOOL

Make a sketch of the **PROFILE** below in the **Front Plane** and make a **BASE - REVOLVE**. Insert an Isometric view on a Title Block and title it **"CABLE SPOOL."**

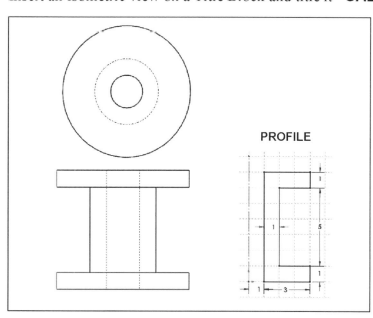

PROFILE

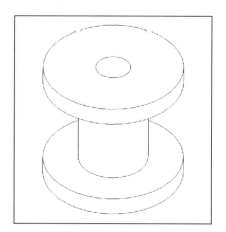

ISOMETRIC VIEW OF THE CABLE SPOOL

ASSUME THE GRID DIVISIONS TO BE 1.00 INCH.

NOTES:

Computer Graphics Lab 2:
2-D Computer Sketching II

ADVANCED 2-D SKETCHING

In the first Computer Graphics Lab, you used some of the basic 2-D sketching capabilities of SOLIDWORKS. These first exercises concentrated on using items that were available on the sketch toolbar. You learned how to draw a Line, Circle, Rectangle, Arc, Polygon, Centerline, and Spline. You also learned how to edit the 2-D sketch using Dimensions, Trim, Mirror, Fillet, and Chamfer functions. In this Computer Graphics Lab 2, you will learn some more advanced 2-D sketching and editing features available in SOLIDWORKS.

SKETCH ENTITIES MENU

The sketch entities shown under the sketch tab are not the only ones available. Many of the icons have a small down arrow next to them. Each of these icons have additional options available for your use. These entities are also accessible under the **Tools – Sketch Entities** and are shown in **Figure 2-1**. Here you can find the following 2-D sketch entities:

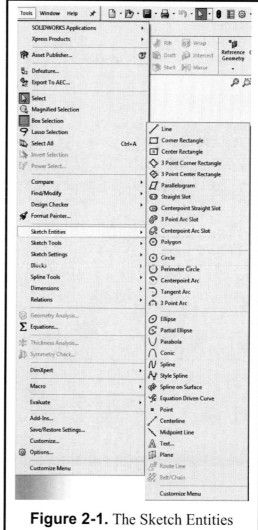

Line
Rectangle (several options)
Parallelogram
Slot (several options)
Polygon
Circle
Perimeter Circle
Centerpoint Arc
Tangent Arc
3 Point Arc
Ellipse (several options)
Partial Ellipse
Parabola
Spline
Spline on Surface
Point
Centerline
Text

Figure 2-1. The Sketch Entities Menu.

Some of these 2-D entities are more commonly used in engineering design than others, but hopefully you will have a chance to use each of them somewhere in one of your exercises.

SKETCH TOOLS MENU

All of the 2-D sketch editing commands are found under the **Sketch Tab**. In this toolbar you will find the following commands:

Fillet is used to round a corner with a radius.

Chamfer is used to cut a corner at an angle.

Offset Entities is used to create a copy of a sketch entity at an offset distance from the original.

Convert Entities projects a 3D feature's edge as a sketch entity in the current sketch.

Trim cuts away a piece of the entity.

Extend extends an entity to meet another entity.

Mirror copies a sketch element about a centerline.

Dynamic Mirror dynamically mirrors new sketch elements about a centerline. First select the entity about which to mirror and then draw the sketch entities to mirror.

Jog Line moves a piece of the line up or down in a rectangular shape.

Construction Geometry converts sketch entities to construction geometry and vice versa.

Linear Sketch Pattern creates a rectangular pattern (row X column) of identical entities (see **Figure 2-3**).

Circular Sketch Pattern creates a radial (or polar) pattern of identical entities around a center point (see **Figure 2-4**).

Align is used to align a sketch and a grid point.

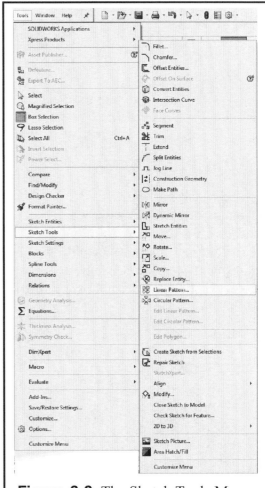

Figure 2-2. The Sketch Tools Menu.

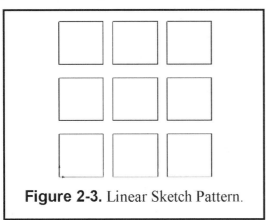

Figure 2-3. Linear Sketch Pattern.

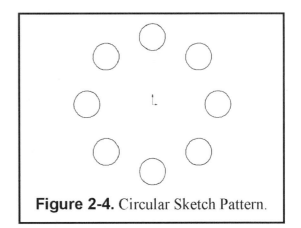

Figure 2-4. Circular Sketch Pattern.

Exercise 2.1: METAL GRATE

In Exercise 2.1, you will design a Metal Grate. The function of a metal grate is such that many identical slots are cut through it. Instead of drawing each slot separately, you will use an advanced sketching command SOLIDWORKS to create a rectangular pattern of slots. Then you can simply extrude the sketch to create the base feature of the grate.

Start by going to your folder and **Open** the file **ANSI-METRIC.prtdot** because the dimensions of the Metal Grate are in Metric units. Immediately **SAVE AS – METAL GRATE.sldprt**. You will not need a grid for this exercise. Go to **Tools – Options - Document Properties** and click the **Grid/Snap** tab and make sure the "Display Grid" function is **not** checked (√) on, then click the **OK** button. Select the **Front** plane in the FeatureManager for the sketch plane.

From the **Sketch Tab** select the **Sketch** command to start. You will first draw two **Rectangles**. The first one is the large outline of the grate and the second one is the initial small rectangular slot that eventually will be patterned. Refer to **Figure 2-5** and use the **Dimension** command to add the sketch dimensions needed. The overall size of the grate is **280** mm by **195** mm, and the (**140** and **97.5**) dimensions are used to center it about the origin. The small slot is **20** mm by **35** mm and is **30** mm below the top and **30** mm to the right of the upper left corner. *Note:* After all the dimensions are applied, the lines turn black. This means that the geometry is fully defined and constrained. Using the fillet command, add **3mm fillets** to the four corners of the small rectangle.

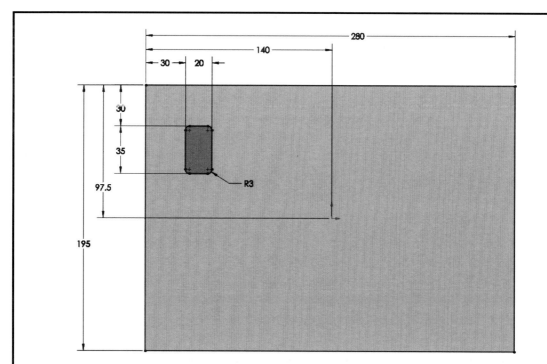

Figure 2-5. The Beginning Dimensions for the Metal Grate Centered at the Origin.

Now select the **Linear Sketch Pattern** command in the Sketch toolbar or the menu **Tools**; select **Sketch Tools** and then click on **Linear Pattern**. The "Linear Pattern Repeat" PropertyManager is automatically displayed. In the Entities to Pattern selection box select the lines and fillets of the small rectangle. The settings for this rectangular Pattern operation are shown in **Figure 2-6** below. "Direction 1" is horizontal and will have **6** instances. The horizontal spacing is **40** mm and the angle is **0** degrees. To activate "Direction 2" change the number of instances to **3**. You will then be able to change the vertical spacing to **50** mm and the angle to **270** degrees. Notice that as you change the options a **Preview** of the pattern is shown. You should now have a 6 x 3 pattern of slots. You will have to select the Reverse Direction option in the Direction 2 section to the left of the Y-axis button to reverse the pattern's direction. Your pattern preview should look like the image in **Figure 2-7**. When finished, click the **OK** button to complete the pattern.

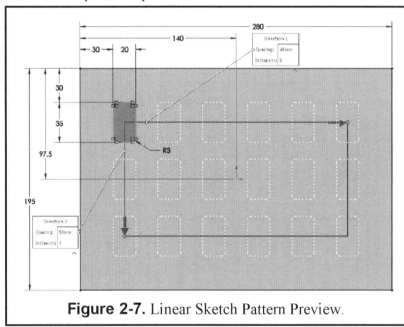

Figure 2-7. Linear Sketch Pattern Preview.

From the **Features Tab** select **Extruded Boss/Base**, enter the following parameters and click OK to finish:

> Type of Extrusion = **Blind**
> Distance 1 = **5** mm

You will now have the base solid model of the grate, as shown in **Figure 2-8** in a **Trimetric** view.

The next step for the Metal Grate is to add a lip to the metal grate in order to provide a support when attached to the wall air duct. Click on the front surface of the Metal Grate. It should highlight *blue*. Then select the **Sketch** tab and click on the **Sketch Command** to add a new sketch. You have already drawn the outer rectangular profile, so you will borrow from it for the outer edge of the lip. From the Sketch tab select the **Convert Entities** command. The outer lines are projected as sketch entities into your active sketch. Notice that they are all black lines since the geometry is already defined.

Figure 2-6. The Linear Sketch Pattern PropertyManager

Convert Entities

Now select the top converted line (it turns *cyan*) and then select the **Offset Entities** command. 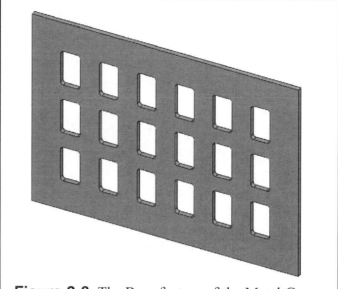 Enter the offset dimension of **15 mm** and make sure the **Select Chain** option box is selected. If the 15 mm offset is previewed on the outside, select the **Reverse** option box so the offset is *inside* of the original lines and then click **OK** to create the offset, as shown in **Figure 2-10**. Also, notice that the offset command places a small 15 mm dimension on your sketch to indicate the offset value. You could now double-click on the dimension, enter in a new dimension value, and change the offset to a new value. But for now leave it at 15 mm. Add a **5mm** Fillet to the inside corners of the offset pattern as shown in **Figure 2-10**.

Figure 2-8. The Base feature of the Metal Grate.

Before you Extrude the sketch, you may want to change to an Isometric view in order to see which direction you are extruding. From the **Features** tab select **Extruded Boss.** In the **Extrude** PropertyManager enter the following parameters:

　　End Condition = **Blind**
　　Distance 1 = **5 mm**

Click OK to complete the boss.

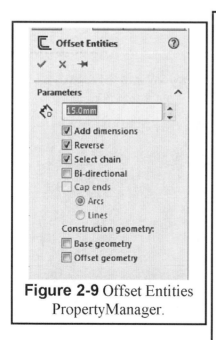

Figure 2-9 Offset Entities PropertyManager.

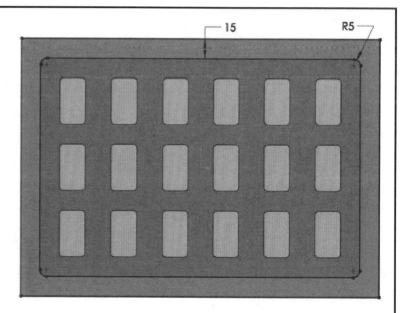

Figure 2-10. Converting the Front Edges, Offsetting them 15 mm and Filleting the Inside Corners with a 5mm radius.

Now you need to add four attachment holes to the corners of the grate. **Select** the face of the last extrusion (it will turn *blue*). From the **Sketch Tab** select the **Sketch** command and draw a **Circle** in the upper left corner. Use the **Dimension** values supplied in **Figure 2-11** for the circle diameter (**8** mm) and position from the corner (**9**mm x **9**mm).

Now draw three more **Circle**s in the other three corners. **Dimension** them to have the same diameter (**8**) and same relative position (**9** x **9**) from each corner. *Or*, now that you are an expert with a rectangular array, use the **Linear Sketch Pattern** command instead. If you use this command, then the horizontal distance of the **2** items is **262** mm and the angle is **0** degrees. The vertical distance of the **2** items is **177**mm and the angle is **270** degrees. *Either way*, when you are finished you should have circles at the four corners and click OK to finish the Linear Pattern.

Change the model orientation to a Trimetric view. From the **Features** tab select **Extruded Cut.** Set the end condition type to **Through all** and click OK to finish. The four corner attachment holes are now created on the grate.

The part is now complete, and you can view the lip feature more clearly by using the **Rotate View** command as shown in **Figure 2-12**.

If you would like to change the color of your model, right click on the model name in the FeatureManager tree and select the Appearances command in the context toolbar. You can then assign any color you wish to the model.

Return to a **Trimetric** View of your part as shown in **Figure 2-12**. You should now save your model. Pull down **File,** select **Save As,**

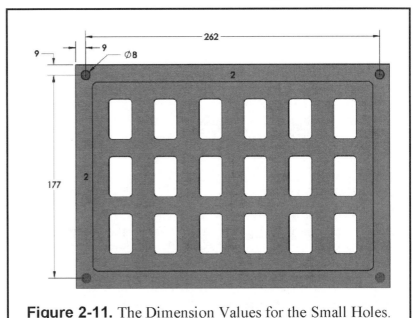

Figure 2-11. The Dimension Values for the Small Holes.

Figure 2-12. The Final Design of the Metal Grate in a Trimetric View.

type in the part name **METAL GRATE.sldprt**, and then click **Save**. Open your copy of **TITLE BLOCK – METRIC.drwdot** and immediately **SAVE AS – METAL GRATE.slddrw**. Now insert the Trimetric Metal Grate view into your **Title Block** template and **Print** it (see **Figure 2-13**).

Print a hard copy to submit to your lab instructor.

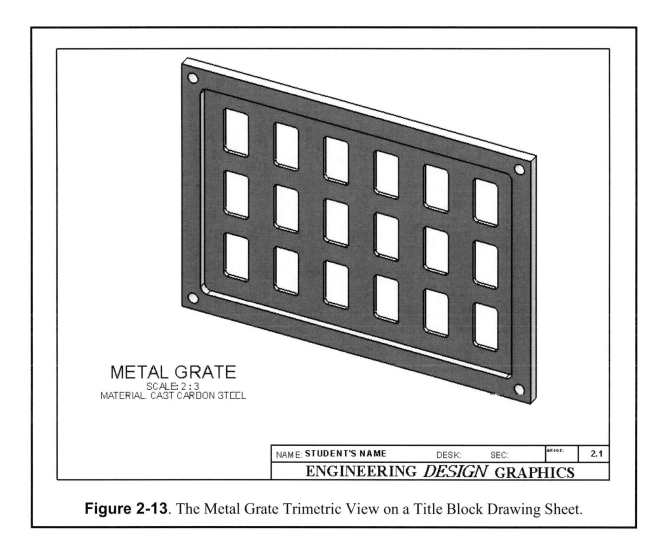

Figure 2-13. The Metal Grate Trimetric View on a Title Block Drawing Sheet.

Exercise 2.2: TORQUE SENSOR

In Exercise 2.2 you will design a Torque Sensor casing. Since it is a cylindrical (axis-symmetrical) object, you will use some of the advanced editing features like circular pattern. Go to your folder and **Open** the file **ANSI-INCHES.prtdot**, and immediately **SAVE AS – TORQUE SENSOR.sldprt**. Select the **Tools**, **Options**, **Document Properties** menus and Select "**Grid/Snap**." Make the following settings on this menu: "**Major grid spacing**" = **1.00**, "**Minor lines per major**" = **4.** Also go to System Snaps and make sure "**Display Grid**" and the "**Snap**" functions are selected, then click the **OK** button. Make sure the **Units** are in **Inches**. Then click **OK.**

The circular features of the Torque Sensor are on the top and bottom surfaces. But the main body is also round and can be created by a 360 degrees solid of revolution using a sketch drawn in the Front plane. Select the **Front** plane in the FeatureManager tree. From the **Sketch Tab** select the **Sketch** command and the sketching grid appears with minor grids spaced every 0.25 inches. Also make sure you are viewing the model from the **Front** view orientation.

First draw a **Centerline** vertically through the origin. Next, use the **Line** tool to sketch the closed sketch shown in **Figure 2-14**. This sketch will generate a part that is 2.50 inches tall and 4.00 inches in diameter at the top and bottom of the part.

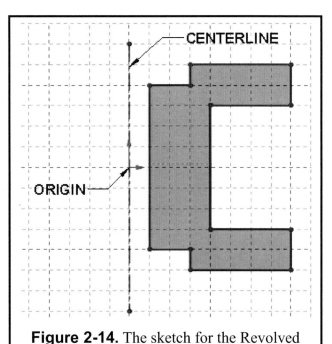

Figure 2-14. The sketch for the Revolved base feature.

From the **Features** tab **Select** the **Revolved Boss/Base** command. Make sure that the centerline is selected for the Axis of Revolution. Enter **360°** to make a full revolution and click OK to complete the revolved boss. The cylindrical base is shown in **Figure 2-15** in an **Isometric** view.

The next step is to create a circular pattern with eight holes around the edge in a bolt circle on the top surface of the part.

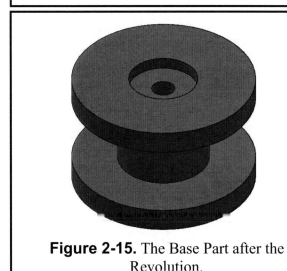

Figure 2-15. The Base Part after the Revolution.

Select the *top* face of the part and it should highlight *blue*. Also change to a **Top** view orientation. Select the **Sketch** command. **Draw** a **Circle** that is **3.25** inches in diameter, and activate the "**for construction**" option in the PropertyManager. Then draw a horizontal construction line from the origin to the right side of the construction circle. The intersection of these two entities defines the center of the first of eight holes, thus resulting in a radius of 1.625. Or you can go to the **Document Properties** menu and on the **Grid/Snap** tab change the "Minor lines per major" value to **8**, thus resulting in a one-eighth inch grid.

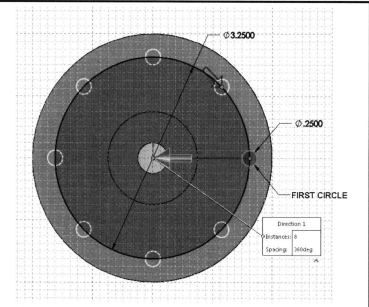

Figure 2-16. Sketching the First Circle and Executing a Circular Array of Eight Holes.

Also on the **Units** tab change the decimal places to **3**. Click **OK** and the grid should now be updated to the new values. Now locate the center of the first **Circle** on the grid and draw it with a diameter of **0.25**. Use **Figure 2-16** to aid you.

Select the circle (it should highlight *blue*). From the drop-down menu in the **Linear Sketch Pattern** select the **Circular Sketch Pattern** command. The "Circular Pattern" PropertyManager is displayed on the left. Referring to **Figure 2-17**, set the parameters for the circular pattern. The "Radius" is **1.625** from the center (**0,0**). The "Step Number" is **8** for a "Total angle" of **360°**. Make sure the "Equal spacing" option is selected, and click **OK** to finish. You now have a bolt circle of 8 holes as shown in **Figure 2-16**. You are now ready to cut these holes through the entire model.

Switch to the **Shaded** model mode and change to an **Isometric** view orientation to better visualize the next operation. From the **Features** tab select the **Extruded Cut** command. In the "Cut Extrude" PropertyManager select the **Through all** end condition and click OK to make the cut extrusion all the way through the model. Use the **Rotate View**

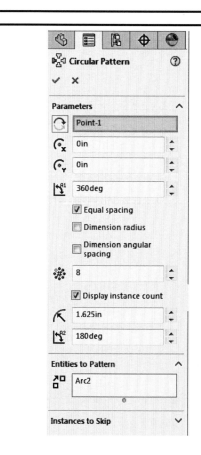

Figure 2-17. The "Circular Pattern" Menu to Create the Bolt Circle Holes.

command to verify the holes were cut all the way through the bottom of the model. The model is complete as shown in **Figure 2-18**.

If you would like to change the color of your model, right click on the model name in the FeatureManager tree and select the Appearances command from the context toolbar to assign any color you wish to the model.

You should now save your model. Pull down **File,** select **Save As,** type in the part name **TORQUE-SENSOR.sldprt**, and then click **Save**. Open your **TITLE BLOCK – INCHES.drwdot** and **SAVE AS: TORQUE SENSOR.slddrw.** Now insert the shaded Isometric view of the Torque Sensor into your **Title Block** drawing sheet that was created in Chapter 1 and **Print** it for your instructor (see **Figure 2-19**).

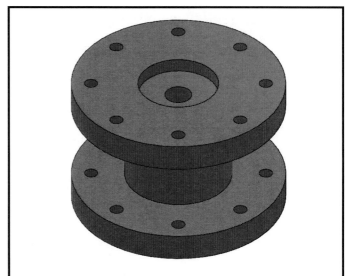

Figure 2-18. The Finished Model of the Torque Sensor in an Isometric View.

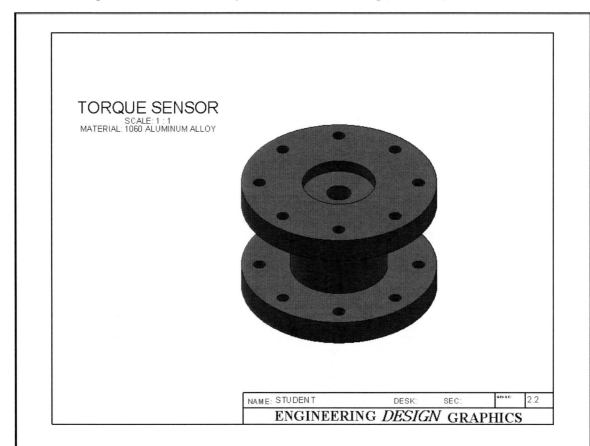

Figure 2-19. The Torque Sensor Shaded Isometric view on a Title Block Drawing Sheet.

Exercise 2.3: SCALLOPED KNOB

In Exercise 2.3, you will design a Scalloped Knob that has some complicated geometry around its edges. This particular knob design will be a hexagon type. Since the hexagonal features are equally spaced around the center of the knob, you can use a circular pattern.

Start by going to your folder and **Open** the file **ANSI-INCHES.prtdot** and immediately **SAVE AS – SCALLOPED KNOB.sldprt**. Go to **TOOLS – OPTIONS – DOCUMENT PROPERTIES** and change the **UNITS** to three decimals. Select the **Front** plane for the sketch. Then start a new **Sketch**. Complete the initial geometry of the sketch according to **Figure 2-20**. Using the **Line** tool, draw two vertical lines and cap them off with a horizontal line that touches their top ends. **Fillet** the top two corners with a **0.10** radius. Use the **Dimension** tool to define the geometry by adding the dimensions shown in **Figure 2-20**. Add the dimensions referenced to the origin. When the geometry is defined, all lines turn *black*.

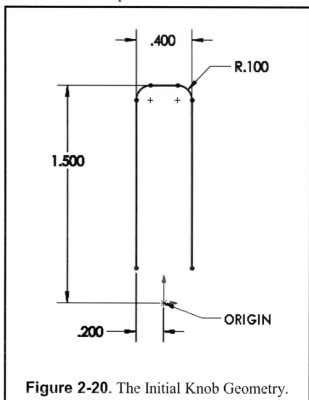

Figure 2-20. The Initial Knob Geometry.

Now make a circular pattern to form a hexagonal layout. From the **Linear Sketch Pattern** drop-down menu select the Circular Sketch Pattern command. When selected you will see the **Circular Sketch Pattern** PropertyManager. Set the "Step Number" to **6** with a "Total angle" of **360°**. Make sure the "Equal spacing" option is selected. Activate the "Entities to Pattern" box and **select the three straight lines and the two fillets** and click **OK** to finish. You now have a pattern that is the first step of the sketch for the knob outline. Notice that some of the lines may overlap as can be seen in **Figure 2-21**. You may want to trim the intersecting lines; however, that is not necessary to complete the remainder of the exercise.

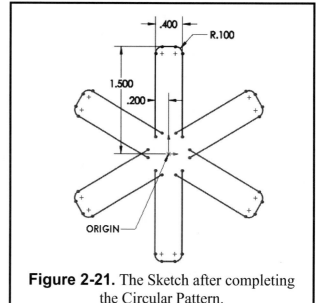

Figure 2-21. The Sketch after completing the Circular Pattern.

Next, you will add a fillet to the six sharp inner corners to create the scallop effect. Select the **Sketch Fillet** command and enter the fillet radius as **0.45** in the PropertyManager. Now select two intersecting lines. A large 0.45 radius is added and a dimension is attached to show the fillet value. Repeat this filleting process on the remaining five sharp inner corners. When you are finished, your sketch should look like **Figure 2-22**.

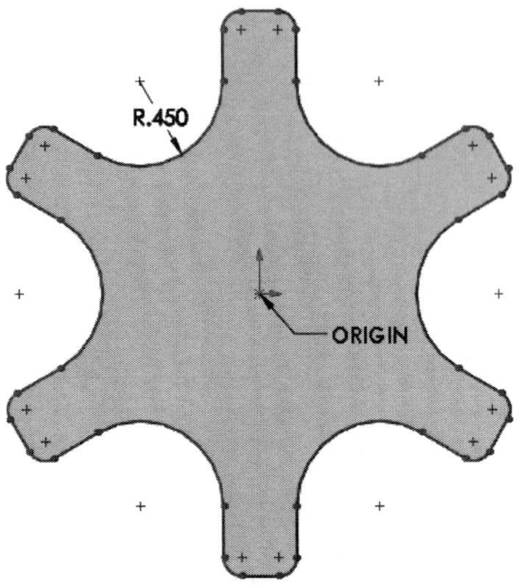

From the **Features** tab select **Extruded Boss/Base**. Extrude the sketch using the **Blind** end condition to a depth of **0.375** inches. Click OK to finish the operation. When finished, change the to a **Trimetric** view orientation as shown in **Figure 2-23**.

Figure 2-22. The Finished Sketch after Filleting Six Sharp Inner Corners.

You now can finish the part by adding the attachment base. **Select the front face** and change to a **Front** view orientation. Select the **Sketch** command and draw a **Circle**, centered at the origin. **Dimension** the circle to be **1.125** inch in diameter. Now draw a **Hexagon** at the origin. Check **Inscribed Circle** and set the diameter to **0.625** inches. From the **Features** tab select the **Extrude** command. Change to an **Isometric view** for better visibility when making an extrusion of any kind. Extrude the sketch to a **Blind** depth of **.75** inches away from the front face.

Change to a **Dimetric** view orientation to see the inside of the hexagonal hole. From the Features tab select the **Fillet** command and set the radius to 0.05". Select the edges of the first extrusion with the 0.45" radius and the edge connecting the first

Figure 2-23. The Extruded Sketch.

extrusion with the second one to remove the sharp edges of the knob and click OK to finish.

If you would like to change the color of your model, right click on the model name in the FeatureManager tree and select the Appearances command from the context toolbar to assign any color you wish to the model.

Now save your model to your designated folder. Pull down **File**, select **Save As**, type in the part name **SCALLOPED KNOB.sldprt**, and then click **Save**. Open your **TITLE BLOCK INCHES.drwdot** and immediately **SAVE AS – SCALLOPED KNOB.slddrw**. Now add a shaded Trimetric view of the Scalloped Knob onto your **Title Block** drawing sheet created in Chapter 1 (see **Figure 2-26**).

Print a hard copy to submit to your lab instructor.

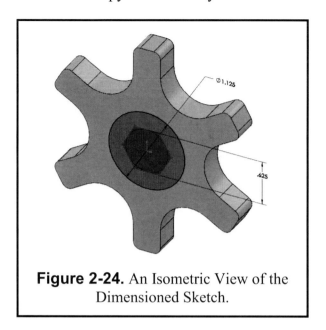

Figure 2-24. An Isometric View of the Dimensioned Sketch.

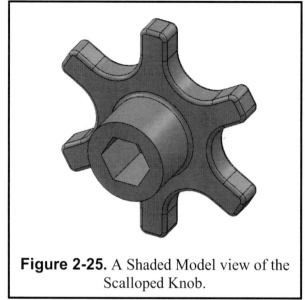

Figure 2-25. A Shaded Model view of the Scalloped Knob.

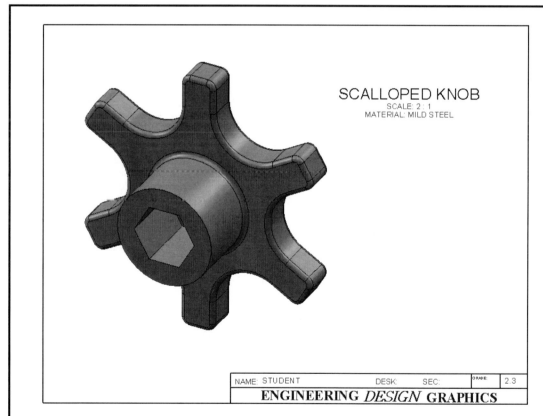

Figure 2-26. The Scalloped Knob Shaded Isometric view on a Title Block Drawing Sheet.

Exercise 2.4: LINEAR STEP PLATE

In Exercise 2.4, you will design a Linear Step Plate used for linear motion control in machinery. There are a lot of holes on this plate, and you will find the linear pattern and mirror commands to be quite helpful. Start by going to your folder and **Open** the file **ANSI-INCHES.prtdot**; immediately **SAVE AS – LINEAR STEP PLATE.sldprt**. Select the **Right Plane,** change to a **Right** view orientation for better visibility and add a new **Sketch**. Draw a vertical centerline through the origin. Go to **TOOLS – Sketch Tools**, and Select **Dynamic Mirror**. Now sketch the right half of the profile shown in **Figure 2-27**. Each line drawn on the right side of the centerline will be duplicated on the left. Use the **Dimension** command to define the geometry by adding the dimensions shown in the **Figure 2-27**.

From the **Features** tab select **Extruded Boss/Base**. On the "Base Extrude" PropertyManager set the extrude parameters as shown in **Figure 2-28**:

Direction 1: **Blind, 4.2000 in.**
Direction 2: **Blind, 4.2000 in.**

OR

Extrude the Sketch **8.4"** using the **MID-PLANE** end condition.

Notice that you can preview this operation in an **Isometric** view on the screen. Click **OK** to complete the extrusion in two directions. The base part looks like **Figure 2-29**.

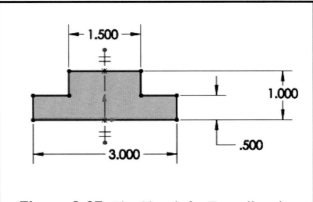

Figure 2-27. The Sketch for Extruding the Base Part.

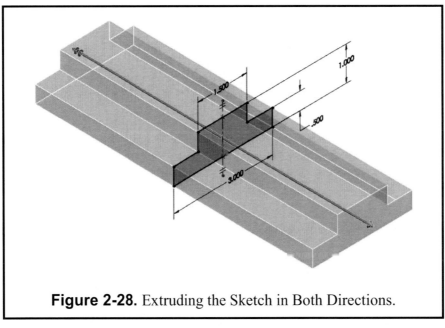

Figure 2-28. Extruding the Sketch in Both Directions.

Now you will create some holes. **Select the top surface of the small step on the front side** (see **Figure 2-29**). **It should highlight** *blue*. In a **Top** view, select the **Sketch** command and draw a small **Circle** on the surface as shown in **Figure 2-30**. Use the **Dimension** tool to add the three dimensions given to define it:

> Diameter = **0.300**
> From center origin = **1.125**
> From center origin = **3.000**

Now create a Linear Sketch Pattern. **Select** the circle (it should turn *cyan*). Select the **Linear Sketch Pattern** from the Sketch tab. The "Linear Pattern" PropertyManager is displayed. Enter the following parameters and click OK:

Direction 1:
> Number = **6**
> Spacing = **1.2000**
> Angle = repeat to right side

Direction 2:
> Number = **1**

You should now have six circles on the front step surface. You need to add six more circles to the back step surface. You can mirror them.

Draw a horizontal **Centerline** across the origin (Note the "—" symbol on your cursor means horizontal). Select the **Mirror** sketch command and the mirror PropertyManager is shown. For the "Entities to Mirror," select the six circles just created with the Linear pattern and in the "Mirror About" box select the centerline drawn through the origin and click OK to finish. The selected items will be mirrored about the centerline, as shown in **Figure 2-31**.

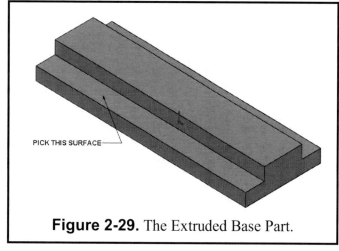

Figure 2-29. The Extruded Base Part.

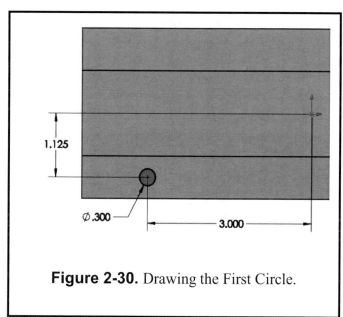

Figure 2-30. Drawing the First Circle.

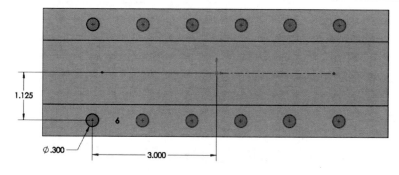

Figure 2-31. Linearly Repeating and Mirroring the Circles.

From the **Features** tab select **Extruded Cut**. Use the **Through All** option and click OK to finish. You have now drilled the small holes all the way through the plate's steps. You now need to bore some counterbore holes a quarter of the way down the small through holes. *Note:* This design feature is called a "Counterbore" and SOLIDWORKS has a

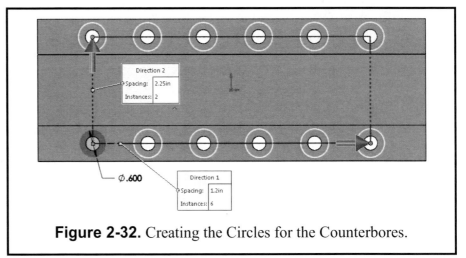

Figure 2-32. Creating the Circles for the Counterbores.

special "Wizard" that can create it. However, we will leave that "Wizard" for a later lab.

Select the top surface of the front step again. **Sketch** a **Circle** on that surface. Then add a relation to make the circle concentric with the hole beneath it (the diameter is **.60**). Now repeat the exact same process as before to get the twelve circles for the counterbore holes.

- o **Select** the new circle.
- o Add a **Linear Pattern** to get the front **6** circles at **1.20** inches apart in **Direction 1**.
- o **In Direction 2** increase the instances to **2** at a distance of **2.25 inches**.

From the **Features** tab select **Extruded Cut**. Use the **Blind** option to a depth of **0.125** inches into the material and click OK to finish. You now bored the counterbores into the plate's two steps, as shown in **Figure 2-33** in a **Rotated View**.

Note: Sometimes you might make a mistake with a **FEATURE** operation like this one. You can simply right mouse click on its name in the FeatureManager and select **Edit Feature** from the context menu. See **Figure 2-34**.

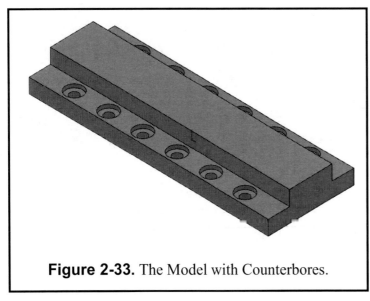

Figure 2-33. The Model with Counterbores.

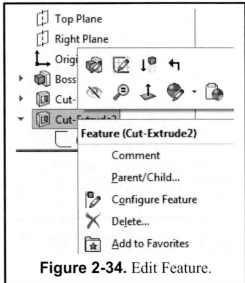

Figure 2-34. Edit Feature.

The next design requirement is to create four holes on the top of the plate. Change to a **Top** view and select the top surface to **Sketch** on. Draw the first **Circle** with the **Dimension** values given in **Figure 2-35**. Use a **Linear Pattern** operation to get a second circle **1.2000** inches from the first circle. Draw a vertical **Centerline** through the origin

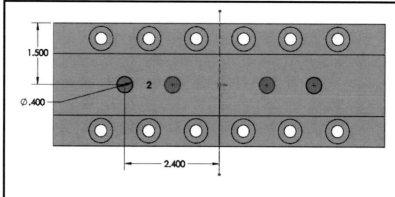

Figure 2-35. Creating the Holes in the Top Surface.

(a | appears on the cursor). Then **Mirror** the two circles. This results in four circles as shown in **Figure 2-35**.

Change **to an Isometric view** for better visibility. From the **Features** tab select **Extruded Cut**. Use the **Through All** option and OK to finish. You now drilled the small holes all the way through the thick part of the plate. **Select the top surface** again and begin a new sketch. You will now add two counterbore slots. Select the **Slot** command from the sketch tab and use the **straight slot** type. Select the center of the left circle on the top face and the one immediately to its right. This will make the slot concentric with the two circles to the left of the center. You can use an identical process to sketch the slot to the right of the center. Dimension the arcs of both slots to have a **Radius of .40**. Then add a **Cut Extrude** feature with a **Blind** depth of **0.250**. These counterbore slots are shown in **Figure 2-37**. To finish the step plate add a chamfer to the three horizontal edges on both ends of the model. From the **Features tab** and under the pull-down **Fillet** command, **Select Chamfer**. Set the chamfer value to **.125**, then select the three top horizontal edges of the ends of the

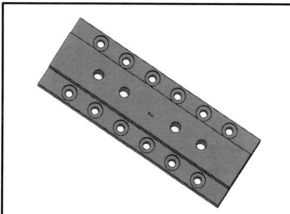

Figure 2-36. The Four Through Holes in the Top Surface.

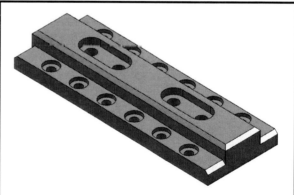

Figure 2-37. The Completed Linear Step Plate.

step plate and the two long edges of the top surface. Click OK to complete the exercise.

In the FeatureManager Tree, Right click on **Edit Material**, expand the **Copper Alloy** materials category and assign **Brass** to the Linear Step Plate.

The part is now finished. Return to an **Isometric** view of the finished part as shown in **Figure 2-37**. Pull down **File,** select **Save As,** type in the part name **LINEAR STEP PLATE.sldprt,** and then click **Save.** Open your **TITLE BLOCK – INCHES.drwdot** and immediately **SAVE AS – LINEAR STEP PLATE.slddrw.** Now insert the shaded Isometric Linear Step Plate view onto your **Title Block** drawing sheet created in Chapter 1 and **Print** it (see **Figure 2-38**).

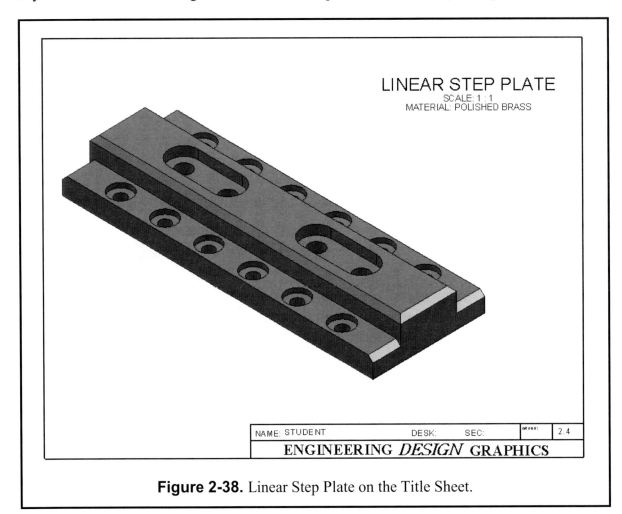

LINEAR STEP PLATE
SCALE: 1 : 1
MATERIAL: POLISHED BRASS

NAME: STUDENT   DESK: SEC: DRAWING: 2.4

ENGINEERING *DESIGN* GRAPHICS

Figure 2-38. Linear Step Plate on the Title Sheet.

SUPPLEMENTARY EXERCISE 2-5: FLANGE

Build the following model using the **Revolve** and **Circular Pattern** commands learned in Unit 2. Add the sketch for the **Revolve** feature in the **Top Plane** using the grid divisions as a guide. Add it on a Title Block drawing and name it **"FLANGE."**

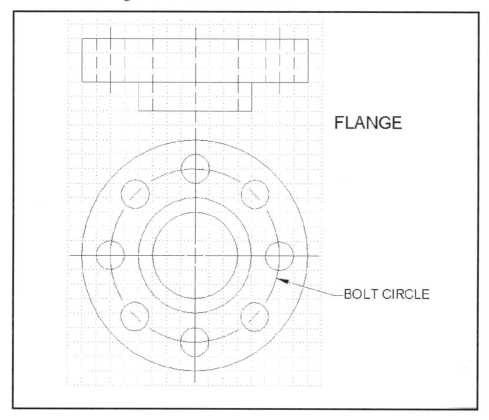

ASSUME THE GRID DIVISIONS TO BE 0.50 INCHES.

SUPPLEMENTARY EXERCISE 2-6:
STEEL VISE BASE

Make a model of the figure below using the commands learned in Unit 2. Insert it onto a Title Block drawing and name it **STEEL VISE BASE**.

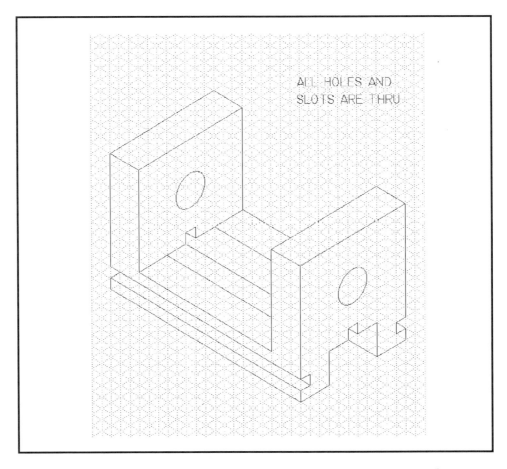

ALL HOLES AND
SLOTS ARE THRU

ASSUME THE GRID DIVISIONS TO BE 0.25 INCHES.

Computer Graphics Lab 3: 3-D Solid Modeling of Parts I

In your first two Computer Graphics Labs #1 and #2, you created some simple solid parts by using 2-D sketches. Almost all of the geometric size and shape data were defined in the 2-D sketch and you simply extruded or revolved the sketch to get the 3D solid model. While this is the normal beginning procedure, SOLIDWORKS contains a variety of advanced commands in which your geometric data can be created or edited directly in 3D space. In this Computer Graphics Lab #3, you will start to create solid models using both 2D sketches and 3D features.

ADDING SKETCH RELATIONS

The 2D sketch will continue to be the normal to start the construction of a solid model. You have already learned how to draw a variety of sketch entities, and how to edit and dimension them. One more capability in the 2D sketch to better define the geometry is to **Add Relations**. The Relations toolbar is shown in **Figure 3-1** and can be placed on screen by going to **View – Toolbars** and selecting **Dimensions/Relations**. Relations define the location of one or more sketch entities. The following are the common relations you may find useful in making a 2D sketch.

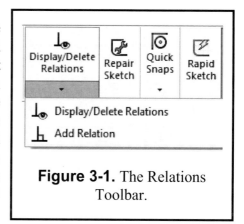

Figure 3-1. The Relations Toolbar.

Horizontal makes one or more lines horizontal, or two or more endpoints horizontal.

Vertical makes one or more lines vertical, or two or more endpoints vertical.

Collinear makes two or more lines lie on the same infinite line.

Co-radial makes two or more arcs share the same center point and radius.

Perpendicular makes two lines perpendicular to each other.

Parallel makes two or more lines parallel to each other.

Tangent makes an arc, ellipse, or spline, and a line or arc tangent to each other.

Concentric makes two or more arcs, or a point and an arc to share the same center point.

Midpoint makes an endpoint of one line to remain at the midpoint of another line.

Intersection makes two lines and one point to remain at the intersection of the lines.

Coincident makes a point and a line, arc, or ellipse to lie on the line, arc, or ellipse.

Equal makes two or more lines or two or more arcs have the same lengths or radii.

Symmetric makes a centerline and two points, lines, arcs, or ellipses to be equal and symmetric about the centerline.

Fix makes the entity's size and location fixed.

THE FEATURES TOOLBAR

The "Features" toolbar is located on the top of your screen and is shown in **Figure 3-2**. These represent the main tools available in SOLIDWORKS to create and edit 3D features. Below are descriptions of these features. _Note:_ In SOLIDWORKS, the first solid feature that you build is called a base, and after that they are called bosses for that part.

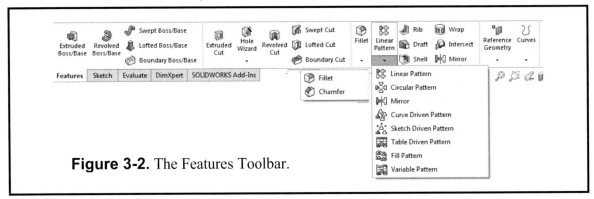

Figure 3-2. The Features Toolbar.

Extrude Boss/Base creates a base or boss by extruding a sketch in a linear direction.

Revolve Boss/Base creates a base or boss by revolving a sketch around a centerline.

Extruded Cut subtracts material from a solid body by linearly extruding a sketch through it.

Revolved Cut subtracts material from a solid body by revolving a sketch around a centerline.

Sweep creates a base, boss, cut, or face by moving a profile along a designated path.

Loft creates a feature by making transitions between multiple profiles.

Fillet creates a rounded internal or external face on the part by selecting an edge.

Chamfer creates a beveled feature on selected edges or a vertex.

Rib adds material of a specified thickness determined by a contour in an existing part.

Shell hollows out the part, and leaves open the faces you select.

Draft tapers faces of a part using a specified angle.

Wrap wraps a closed sketch contour onto a face.

Dome adds a dome to a planar or non-planar face.

Hole Wizard allows you to quickly add different types of standard holes to your part.

Linear Pattern creates multiple instances of selected features along one or two linear directions.

Circular Pattern creates multiple instances of one or more features uniformly around an axis.

Mirror Feature creates a mirror copy of one or more features about a plane.

Reference Geometry creates reference geometry like planes, axis, coordinate systems, and points.

Curves - creates different types of curves including spiral and helix.

Exercise 3.1: CLEVIS MOUNTING BRACKET

In Exercise 3.1, you will design a Clevis Mounting Bracket. You will start with a 2D sketch of the main outline in a front view. Because of its design nature, certain features of the Bracket must remain fixed while other features and dimensions can be varied to accommodate design changes. For example, a hole must remain concentric with an outer arc, but the height of the hole from the bottom base could vary. So you will need to add some geometric relations to the sketch that will define or "constrain" the geometry. In addition, in this exercise you will use some 3D editing commands, like "Mirror Feature," to complete the design.

Go to your folder and **Open** the file **ANSI-INCHES.prtdot**. In order to avoid corrupting the **ANSI-INCHES.prtdot template**, go to **File - Save as:** under **File Name**, type **CLEVIS MOUNTING BRACKET**, and under **Save as type**, select **.sldprt** and select **SAVE**. Select the **Right Plane** in the FeatureManager (it turns *blue*). The Clevis Bracket is symmetric about the right plane in 3D space. So the initial Sketch1 needs to be constructed on a plane that is parallel to the right plane but a distance from it. To better see this operation, click to **Isometric** view orientation. Pull down **Insert**, select **Reference Geometry,** and then select **Plane**. The "Plane" PropertyManager appears as shown in **Figure 3-3**. Indicate the distance, which should be **1.375** inch, then click OK to add the new plane.

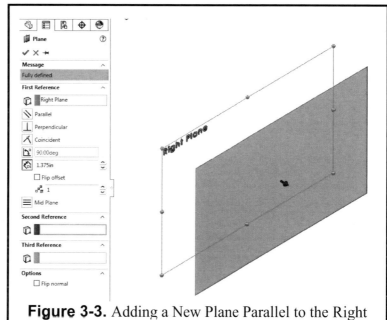

Figure 3-3. Adding a New Plane Parallel to the Right Plane and 1.375 Inch from It.

You now have a new plane in the FeatureManager tree called **Plane1**. Select on this new plane and from the **Sketch Tab** select the **Sketch** command to add a 2-D sketch. Use the **Line**, **Circle**, and **Arc** sketch tools to draw the rough 2D profile shown in **Figure 3-4**. You do not have to be very accurate right now with your sketch because you will be adding geometric relations and dimensions that will constrain the geometry. Just make sure the sketch is a little above the origin as shown in **Figure 3-4**.

In the Sketch tab, click on the **Display/Delete Relations** pull down menu and select the **Add Relations** command (it is the command with the perpendicular symbol). The Add Relations PropertyManager is automatically displayed as shown in **Figure 3-5**. Select the bottom line and it turns *cyan*. It has three sections:

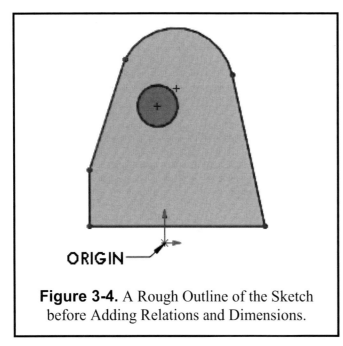

Figure 3-4. A Rough Outline of the Sketch before Adding Relations and Dimensions.

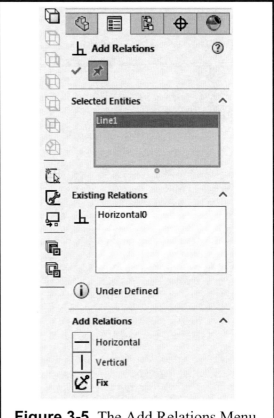

Figure 3-5. The Add Relations Menu.

"Selected Entities" displays the currently selected entity(ies);

"Existing Relations" displays existing relations of the selected item(s);

"Add Relations" lists all the possible relations

the item(s) can have.

Add the **Horizontal** relation to this bottom line and click OK. The bottom line is now horizontal. *Note:* Depending on how you drew the first line, it may already be horizontal because the smart cursor sometimes infers the intent of the designer.

Now study all the remaining relations to add to the sketch, as shown in **Figure 3-6**. Each time select the **Add Relations** command, select the entity(ies) to turn *cyan*, add the appropriate relation(s) in the "Add Relations" PropertyManager, and then click OK to finish.

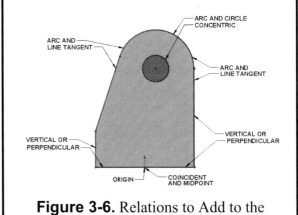

Figure 3-6. Relations to Add to the Sketch.

Repeat these steps to add the following as indicated in **Figure 3-6**.

- Origin and Bottom Horizontal Line are **Coincident** and **Midpoint**. (Adding a **Midpoint** relation also makes it **Coincident**).

- Right Side line is **Vertical** or **Perpendicular** to the bottom line.

- Right Side line and Top Arc are **Tangent**.

- Top Arc and Circle are **Concentric**.

- Top Arc and Left Angled Line are **Tangent**.

This completes all the geometric relations needed in the sketch. However, your current sketch may not look like the final version. It may be too short or too tall or too wide. Also, some lines are *blue* meaning the geometry is under-defined. Try something now. With the **Select** cursor, click and drag the center of the concentric circle and arc and move it up, to the right, or to the left. See the sketch change shape, but the relations (like tangent, vertical, midpoint, and perpendicular) are still maintained. This is called constraint-based modeling. Now return the center point back to its original position.

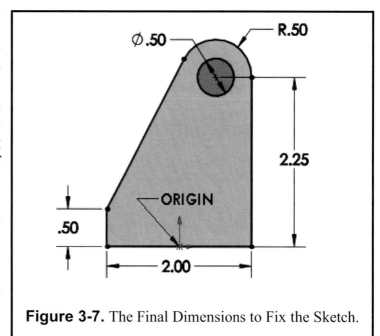

Figure 3-7. The Final Dimensions to Fix the Sketch.

You will now fully define the geometry of your sketch by adding dimensions. Study the dimensions to add in **Figure 3-7**. With the **Dimension** tool, apply the dimensions (all in inches). Start with the **2.25** height of the center of the concentric circle and arc. Then apply a **0.500** diameter and radius dimensions for the circle and arc. Finally, add the **2.000** length of the bottom line and the **0.50** height of the small vertical line on the left side. Now the sketch, with relations and dimensions added, is fully defined and all lines turn *black*.

You can now try another thing. With the **Select** cursor, double click the **2.000** dimension of the bottom line and change it temporarily to **4.000**. Notice how wide the base becomes, but it is centered at the origin because of the "Midpoint" relation we added. Then change that same dimension to **1.000**. See how narrow it becomes, but the "Midpoint" and "Tangent" relations still hold true and stand out in the design. Now return the dimension back to the original value of **2.000**. You are now ready to extrude the sketch to get some 3D geometry for the Clevis Mounting Bracket.

Select the **Extrude Boss/Base** command from the "Features" tab and the "Base-Extrude" Property-Manager is shown. Make a **Blind** extrusion with a distance of **0.375** inches toward the **Origin** and **Right Plane**" (see **Figure 3-8**). You can use the reverse direction button to make the extrusion towards the *Right* plane **(away from your point of view)**. Use an **Isometric** view for better visibility and click OK to finish. You should now have the first 3D feature of the Bracket as shown in **Figure 3-8**.

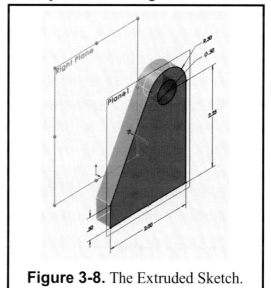

Figure 3-8. The Extruded Sketch.

Now return to a **Right** view and select the front face (it turns *blue*) to sketch on it. From the **Sketch Tab Select** the **Sketch** command. Select the **bottom horizontal edge**, then hold down the **Ctrl** key and select the **left short vertical edge** of the model (they turn *cyan*). Next, click the **Convert Entities** command to convert *those edges* into sketch lines for the current sketch. They are *black* lines fully defined and still reference the model edges from which they were projected. Now draw a **Line**, from the top of the converted vertical line, horizontally over to the right edge. Then draw a final **Line** vertically down to the bottom right corner. This essentially makes a rectangular sketch that can be extruded to form the bottom of the Clevis Mounting Bracket. Select the **Extrude Boss/Base** command from the "Features" tab. Make a **Blind** extrusion with a distance of **2.750** inches <u>back</u> into the direction of the "**Right Plane**" (use Reverse Direction if needed). Change to an **Isometric** view for better visibility, and click OK to finish the bottom base.

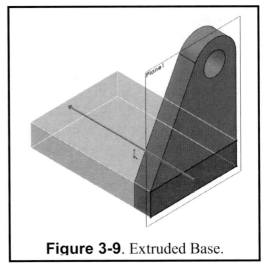

Figure 3-9. Extruded Base.

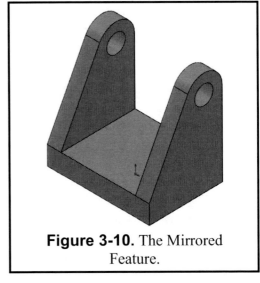

Now you need to create a symmetric feature like **Figure 3-8** on the backside of the model. Hold down the **Ctrl** key, and on the FeatureManager tree, select both the **Right Plane** and the **Boss-Extrude 1** feature. Now select the **Mirror Feature** command on the "Features" tab (see command to right) and click OK to finish. The "Boss-Extrude1" feature is now mirrored about the front plane, as shown in **Figure 3-10**.

⊩⊣ Mirror
Mirror
Feature

Figure 3-10. The Mirrored Feature.

You will now complete the Bracket by adding some features to the bottom base. Refer to **Figure 3-11** for this data. Change to a **Top** view orientation. Select the top face of the bottom base and select the **Sketch** command. Use the **Circle**, **Arc** and **Line** tools to draw the hole and slot outline. Use the **Dimension** tool to add the given dimensions. The slot at the bottom can extend over the edge a little to make sure it cuts off the solid middle part of that edge. **Add a relation so the center point of the circle and the arc are aligned vertically with the origin**. If it is correct, the lines should turn *black* once all dimensions are applied, except for the portion extending beyond the edge.

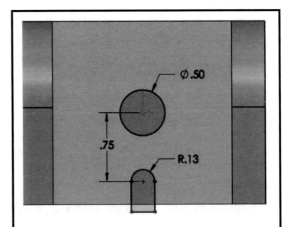

Figure 3-11. Geometry for the Bottom Holes.

Now select the **Extruded Cut** command. In the "Extruded-Cut" PropertyManager, make it **Through All** end condition and make sure the cut direction is downward. Use an **Isometric** view for visibility and click OK to finish. One last operation to complete the model is to add a **0.125 Fillet** to the edges where the vertical faces intersect the horizontal face. Your Clevis Mounting Bracket is now complete. View your final model in an **Isometric** orientation as shown in **Figure 3-12**. The **CLEVIS MOUNTING BRACKET** will be made of Bronze. Change the part's material. In the FeatureManager Tree, right click on **Material, select Edit Material**, expand the **Copper Alloy** materials category and assign **Leaded Commercial Bronze** to the **CLEVIS MOUNTING BRACKET**.

From the **File** menu, select **Save As,** select the proper folder, type in the part name **CLEVIS MOUNTING BRACKET.sldprt**, and then click **Save**. Add an isometric shaded view of the Clevis Mounting Bracket onto your **Title Block** drawing sheet, as shown in **Figure 3-12**. Now **SAVE** your drawing to your directory as **CLEVIS MOUNTING BRACKET.slddrw**. Take note that the title of the part and the drawing are the same but the extension has changed. The solid model and the drawing are linked, meaning that any changes made to the solid model will automatically be updated on the drawing (see **Figure 3-12**).

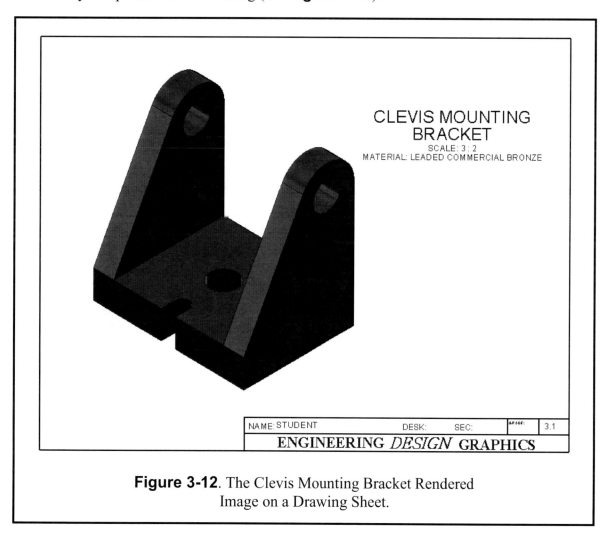

CLEVIS MOUNTING
BRACKET
SCALE: 3 : 2
MATERIAL: LEADED COMMERCIAL BRONZE

| NAME: STUDENT | | DESK: | SEC: | DR ODE: | 3.1 |

ENGINEERING *DESIGN* GRAPHICS

Figure 3-12. The Clevis Mounting Bracket Rendered
Image on a Drawing Sheet.

Exercise 3.2: MANIFOLD

In this Exercise 3.2, you will design a Manifold. The Manifold is designed to allow air or a fluid to flow in many directions through its ports. There will be many similar flow ports, so you will use some 3D editing commands in SOLIDWORKS to replicate them.

Go to your folder and **Open** the file **ANSI-METRIC.prtdot** since the Manifold is designed in metric units. In order to avoid corrupting the **ANSI-METRIC.prtdot** template, go to **File - Save as:** under **File Name**, type **MANIFOLD**, and under **Save as type**, select **.sldprt** and select **SAVE**. Select the **Right** plane in the FeatureManager, select a **Right** view orientation and from the **Sketch** tab select the **Sketch** command to add a new sketch. Draw two concentric **Circles** centered at the **ORIGIN**, with one circle a little larger. **Dimension** the larger circle to be **60** mm in **diameter** and the smaller circle to have a **45** mm **diameter**. These two dimensions will fully define your geometry.

From the **Features** tab select **Extrude Boss/Base**. In the PropertyManager select **Mid Plane** in **Direction 1** and make it **300 mm**, then click OK. You now have a long throat centered at the origin, as shown in **Figure 3-13**.

Figure 3-13. The Base Throat Hole.

Now add a collar on the visible end. Select the right side face of the throat hole (defined by the two concentric circles). The face should turn *blue*. Select the **Sketch Tab** and click on the **Sketch command**. Select the smaller inner circular edge. Select the **Convert Entities** command to project it to the new sketch. Now draw one **Circle**, centered at the origin, and **Dimension** it with a diameter of **75** mm.

Select the **Extrude Boss/Base** command. In the **Direction 1** section of the PropertyManager, select **Blind**, reverse the direction **(back over the throat),** make it **50** mm and click OK to finish. You now have a small collar on one end of the throat. You can easily copy it to the other side. Hold down the **Ctrl** key. In the FeatureManager, select the **Right** plane and the previous **Boss-Extrude2**. Both features will be highlighted in the screen. Activate the **Features** tab, select the **Mirror**

Figure 3-14. Adding Collars to the Ends.

command and click OK to finish the mirror. You now have a collar on each end of the throat, as shown in **Figure 3-15**.

You will design the first port hole. Select the **Top** plane in the FeatureManager. From the **Insert** menu select **Reference Geometry,** and then select **Plane**. In the PropertyManager set the distance to be **42.5** mm above the top plane, and click Ok to add the new plane to the model. Select **Plane1** in the FeatureManager (it turns *cyan*) and start a new **Sketch**. Change to a **Top** view orientation to better see this new sketching plane.

Now draw a small **Circle** on the plane just inside the collar on the left side of the throat. Select the **Add Relations** command, and select both the center of the circle and the origin. In the PropertyManager select **Horizontal** to add a relation to these two entities. Click OK to add this relation to align them horizontally. Next, **Dimension** the circle to have a diameter of **35** mm and to be **75** mm to the left of the origin, as shown in **Figure 3-15**.

Select the **Extrude Boss/Base** command in the features tab. In the PropertyManager set Direction 1 to be **Up to Face**. Select the outer face of the main body of the model (it turns *pink*), then click OK. You now have a boss extruded down from Plane 1 to the outer face.

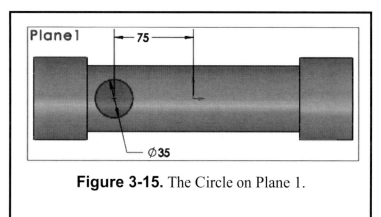

Figure 3-15. The Circle on Plane 1.

Now repeat a **Sketch** on **Plane 1**, and draw a smaller **Circle** that is concentric with the previous boss (**Add Relations, Concentric**) and add a diameter of **17.5** mm.

Return to an **Isometric** view orientation. Select the **Extruded Cut** command on the Features Tab. In the PropertyManager set Direction 1 to be **Up to Face**. Select the *inner throat face* on the model (it turns *pink*) and click OK to finish. You now have a hole going down from Plane 1 into the inside of the throat as shown in **Figure 3-16**.

To better see the through hole, use the **Rotate View** command to rotate your model around the screen. Try to view the through hole by peering down the throat of the part. Later in this exercise you will use a temporary section view to better see the inside.

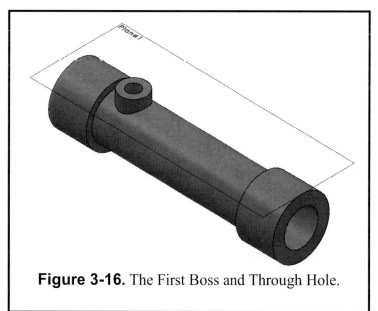

Figure 3-16. The First Boss and Through Hole.

Now that you have one boss and hole feature, it is easy to create a 3D pattern to complete the three that are needed. First you need an axis for the direction of the 3D pattern. Select the menu **Insert**, select **Reference Geometry,** and then select **Axis**. The "Axis" Property Manager is shown in **Figure 3-17**. Select the **Cylindrical/Conical Face** definition for the axis and then click the outer face of the main cylinder. It gets added to the "Selections" list and click OK to finish the axis. You now have added "**Axis1**" to the FeatureManager tree that goes down the center of the throat. Press down the **Ctrl** key, and one-by-one select in the FeatureManager tree:

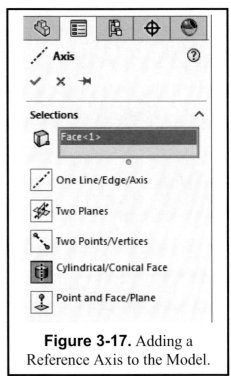

Figure 3-17. Adding a Reference Axis to the Model.

Axis 1
Boss-Extrude 3
Cut-Extrude 1

Now select the **Linear Pattern** command in the **Features tab**. The "Linear Pattern" PropertyManager is shown in **3-18**. There is only one Direction for this case. The parameters for "Direction 1" should be along **Axis 1**, distance of **50** mm, and **4** copies. Use an Isometric orientation to see the copies over the Manifold to the right side. Click OK to finish. You should now have a pattern of four port holes as shown in **Figure 3-19**. Try viewing down the middle of the throat with the **Rotate View** command.

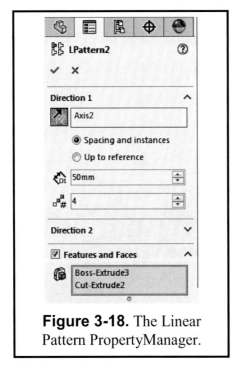

Figure 3-18. The Linear Pattern PropertyManager.

Figure 3-19. A Linear Pattern of Four Port holes.

There are two remaining port holes that need to be added to the bottom of the Manifold. They are identical in geometry to the top, so you can just mirror the port hole about the Top plane. Hold down the **Ctrl** key and, in the FeatureManager tree, one-by-one select **Top Plane**, **Boss-Extrude 3**, and **Cut-Extrude1**. Then click the **Mirror Feature** command and click OK to finish the command. You now have one port hole on the bottom of the Manifold, which is named "Mirror2." Repeat this 3D **Mirror Feature** operation, this time using the **Right Plane** and

Mirror 2. After you complete this operation, you will have a "Mirror3" feature for the Manifold, as shown in **Figure 3-20**.

Now add a fillet feature to the edge where the portholes touch the main body. Select the **Fillet** command on the **FEATURES TAB**. Set the fillet radius **6 mm**. One-by-one select the edges (loops) of the six port holes where they

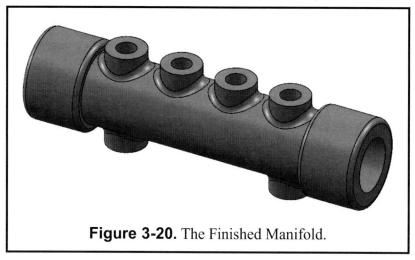

Figure 3-20. The Finished Manifold.

touch the outer face of the throat (they turn cyan). You can do this easily with a **Front** view orientation. Then click OK to finish the fillet command.

Now add a **4mm** radius **Fillet** to the sharp edges of the two collars. See the finished model in **Figure 3-20** with a **Trimetric** view orientation.

Earlier you rotated the model around to see through the long throat hole to see if the port holes went through to the middle. Now you will create a temporary section view to see internal features. From the **View** menu select **Display**, and then select **Section View**. The "Section View" PropertyManager appears on the screen. The section view parameters should be set to **0** mm distance and the **Front** plane is the default plane, as shown in **Figure 3-21**. Just press cancel (**X**) after you have inspected the inside features of the manifold.

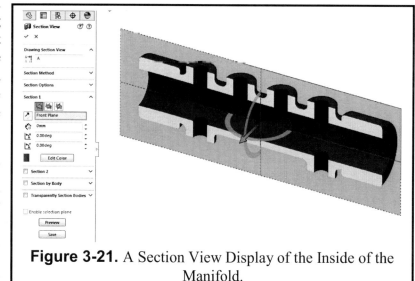

Figure 3-21. A Section View Display of the Inside of the Manifold.

Now you need to change the **MANIFOLD** material to TITANIUM. In the FeatureManager Tree, right click on **Edit Material**, expand the **OTHER METALS** materials category and assign **TITANIUM** to the **MANIFOLD.**

From the **File** menu select **Save As,** type in the part name **MANIFOLD.sldprt**, select your proper folder, and then click **Save**. Add the isometric shaded view into the **Title Block** drawing sheet like in previous labs (refer to page 1.7).

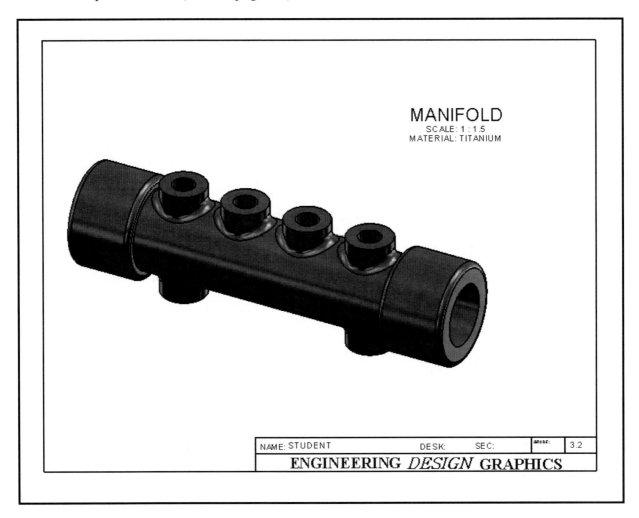

MANIFOLD
SCALE: 1 : 1.5
MATERIAL: TITANIUM

| NAME: STUDENT | DESK: | SEC: | GRADE: | 3.2 |

ENGINEERING *DESIGN* GRAPHICS

Exercise 3.3: HAND WHEEL

In this Exercise 3.3, you will design a Hand Wheel. The Hand Wheel has an elliptical cross-section that can be revolved 360 degrees. You will learn how to sketch an ellipse. The five spokes for the hand wheel can be created using a 3-D circular pattern. You will employ some new SOLIDWORKS commands in this lab exercise.

Go to your folder and **Open** the file **ANSI-INCHES.prtdot.** Immediately go to **File - Save as:** under **File Name**, type **HAND WHEEL**, and under **Save as type**, select **.sldprt** and select **SAVE**. Select the **Front** plane in the FeatureManager tree. Set the view orientation to **Front**, then start a new **Sketch**. If you do not have an ellipse command, you need to go to **Tools**, select **Sketch Entity**, and then select **Ellipse**. Draw an ellipse off to the right side of the origin as shown in **Figure 3-22**. Click to place the center of the ellipse on to the right side of the origin. Click again **vertically** above the center to set the major axis, and finally click to the right to set the minor axis of the ellipse.

Right now the ellipse is randomly placed on the Front sketch plane. The center of the ellipse should be aligned with the origin. Click on the **Display/Delete Relations down arrow** and select the **Add Relations** command. Select the **center of the ellipse** and then select the **Origin**. Add the **Horizontal** relation to align them and then click OK. Now **Dimension** the ellipse using **Figure 3-22** as a guide. The major diameter is **1.000** inches, the minor diameter is **0.750** inches, and the center's distance from the origin is **4.500** inches. Also sketch a **vertical Centerline** through the origin as shown in **Figure 3-22**. The ellipse is now ready to be revolved.

Select the **Revolve Boss/Base** command from the features toolbar. In the PropertyManager, set the parameters to **One Direction** and **360** degrees, then click OK. You now have the first feature for the Hand Wheel as shown in **Figure 3-23**

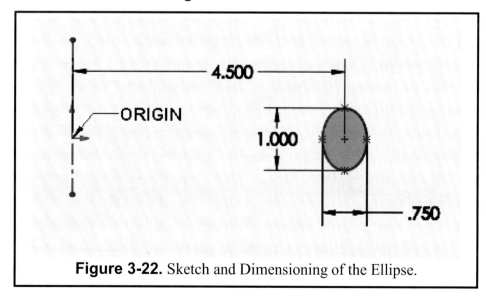

Figure 3-22. Sketch and Dimensioning of the Ellipse.

You will now model a spoke from the center of the Hand Wheel. It also has an elliptical cross section as shown in **Figure 3-24**. Select the **Front Plane** in the FeatureManager and change to a **Front** view orientation. **Sketch** an **Ellipse** centered at the origin. Next, **Dimension** the ellipse according to **Figure 3-24**.

Switch to an **Isometric** view orientation to better see the next operation. Select the **Extrude Boss/Base** command and select the **Up to Surface** end condition for Direction 1. Now **Select** the inside face of the Hand Wheel (**Figure 3-23**) to define the face for the end condition for the spoke. It should turn *pink* and the extruded spoke preview will be visible on the screen. If it looks correct, click OK to finish. You should now have one spoke as shown in **Figure 3-25**.

You can now add the remaining spokes using a circular pattern feature. But first you need to add an axis for the pattern function. If the origin is not visible, pull down the **View – Hide/Show** menu and select **Origins**. Select the **Insert**, **Reference Geometry**, and select **Axis**. The "Reference Axis" PropertyManager appears on the screen as shown in **Figure 3-26**. Now **Select Two Planes** as the axis option. Using the fly-out FeatureManager to the right of the PropertyManager, select the **Front Plane** and the **Right Plane**. Then click **OK** to finish. You should now see the Axis1 feature on your screen.

Figure 3-23. The First Feature of the Hand Wheel.

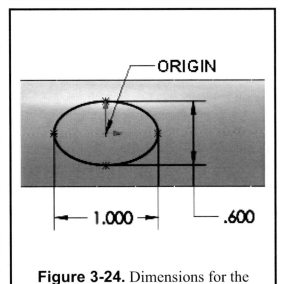

Figure 3-24. Dimensions for the Elliptical Spoke.

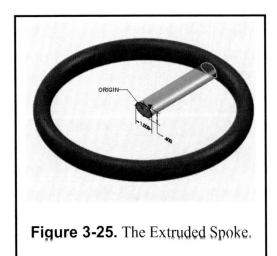

Figure 3-25. The Extruded Spoke.

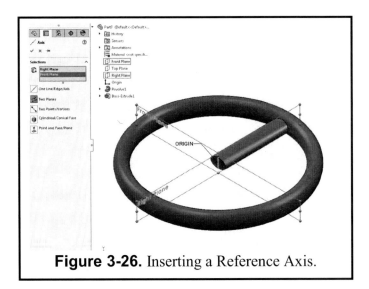

Figure 3-26. Inserting a Reference Axis.

Hold down the **Ctrl** key and **Select** both the **Axis 1** (you can select it in the FeatureManager area) and the **Spoke**. They will highlight in *blue*. Select the **Circular Pattern** command on the Features Toolbar. In the "Circular Pattern" PropertyManager, the axis and extrusions are pre-selected, **Axis 1** should appear in the Pattern Axis selection box and **Boss-Extrude1** should appear in the "Features to Pattern" selection box. Now set the "Angle" to **360** degrees and the number of copies to **5**. Now click OK to finish the circular pattern of spokes, as shown in **Figure 3-27**.

Figure 3-27. Circular Pattern of Spokes.

Now you will add the central hub of the Hand Wheel. Select the **Top** plane in the FeatureManager and change to a **Top** view orientation. From the **Sketch Tab Select** the **Sketch Command** and draw a **Circle** centered at the origin with a diameter **Dimension** of **2.000** inches. Switch to **Isometric** view orientation. Select the **Extrude Boss/Base** command, and enter the following parameters:

Direction 1: **Mid Plane, 1.50** inches

Figure 3-28. The Central Hub Added.

Click OK to finish the central hub. Notice the center boss hub is extruded in both directions as shown in **Figure 3-28**.

You will now add the hole and keyway to go through the boss. Select the top face of the central boss and change to a **Top** view orientation. Now click on the **Sketch** command and draw the sketch shown in **Figure 3-29**. You will need to draw a **Circle** and then either a few **Lines** or maybe a **Rectangle**. Use the **Trim** tool to cut away the profile for the keyway. Then **Dimension** the sketch as shown in **Figure 3-29**:

    Circle Diameter = **1.25** inches
    Keyway width = **0.220** inches
    Keyway height = **0.750** inches
    Keyway distance = **0.110** inches

If all is correct, your sketch should turn *black* meaning the geometry is fully defined.

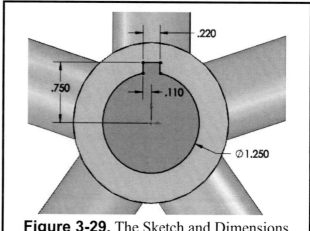

Figure 3-29. The Sketch and Dimensions for the Through Hole and Keyway.

Now click the **Extrude Cut** command and select the **Through All** end condition. Make sure the direction for the "through all" condition is down, then click OK to finish. The cut is created through the hub as shown in **Figure 3-30**. The Hand Wheel is almost complete; you just need to add the fillets to the places where the spokes intersect the hub and wheel hub.

Now click the **Fillet** command on the Features tab (it looks like a block with a round corner). Enter a fillet radius of **0.125** inches. Select the five faces of the five spokes to fillet the edges. They turn *blue* when selected and they get added to the "Items to Fillet" selection list in the menu and click OK to add the ten fillets. The finished Hand Wheel model is shown in **Figure 3-30** in an **Isometric** view.

Figure 3-30. The Finished Hand Wheel Model Shown in Isometric.

Select the **File** menu, select **Save As,** type in the part name **HAND WHEEL.sldprt**, select your proper folder, and then click **Save**. Insert the isometric view into the **Title Block** drawing sheet like in previous labs (see **Figure 3-31**). **Print** a hard copy to submit to your lab instructor. Save your drawing as **HAND WHEEL.slddrw**.

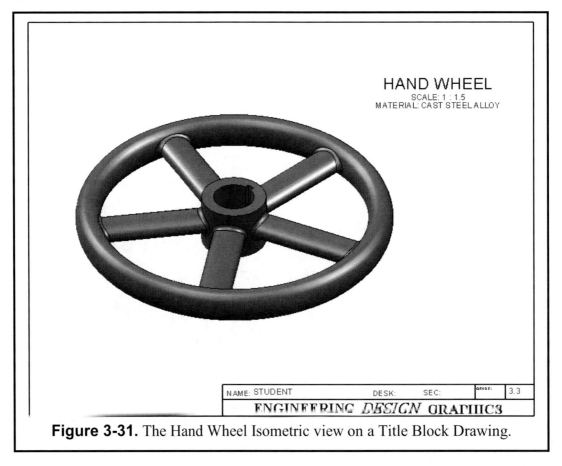

HAND WHEEL
SCALE: 1 : 1.5
MATERIAL: CAST STEEL ALLOY

NAME: STUDENT   DESK: SEC: 3.3

ENGINEERING *DESIGN* GRAPHICS

Figure 3-31. The Hand Wheel Isometric view on a Title Block Drawing.

Exercise 3.4: TOE CLAMP

In the previous lab exercise, you created sketches on planes that were orthogonal to the principal planes (Front, Top, or Right). In this Exercise, you will create features for the Toe Clamp by sketching on both orthogonal and inclined planes. The inclined plane is at a 45 degree angle to the top face, so you will also learn how to dimension angles.

Go to your folder and **Open** the file **ANSI-INCHES.prtdot**. Immediately go to **File - Save as:** under **File Name**, type **TOE CLAMP**, and under **Save as type**, select **.sldprt** and select **SAVE**. Click on the **Front** plane in the FeatureManager tree. Set the view orientation to **Front**, then start a new **Sketch**. You need to draw the sketch of the Toe Clamp as shown in **Figure 3-32**. Use the **Line** and **Dimension** tools to create and fix the geometry. **Add Relations** to the bottom **Horizontal** line and make it both **Coincident** and **Midpoint** with the origin. When dimensioning an angle, select the two lines that form the angle and then click to locate the angle dimension.

Now select the **Extrude Boss/Base** command to create the base part, using the following parameters:

Direction 1:

Mid Plane, **2.00** inches

Click OK to complete the bi-directional base extrude, as shown in **Figure 3-33**.

You will now cut the counterbore and hole on the inclined face. Click on larger inclined face to select it. Now you want to see this face head on (normal to the screen). Select the **Normal to** command in the View Orientation box (or you can also click the **Normal to** command on the top View Orientation toolbar shown below).

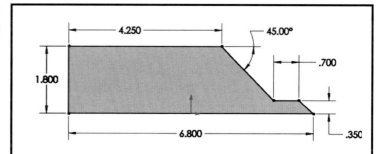

Figure 3-32. The Initial Shape and Dimensions for the Toe Clamp.

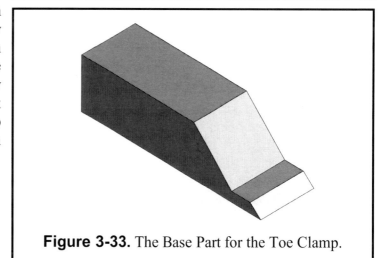

Figure 3-33. The Base Part for the Toe Clamp.

Add a new **Sketch** and draw a **Circle** on the inclined plane as shown in **Figure 3-34**. **Dimension** the diameter to be **0.50** inches. Dimension the center of the circle **1.00** inches vertically. Add a **vertical** geometric relation between the center of the circle and the origin. Select the **Extruded Cut** command on the Features toolbar using the **Through all** end condition. Make sure the hole is cut perpendicular to the angled face, and click OK to finish. You can inspect the model with the **Rotate View** command.

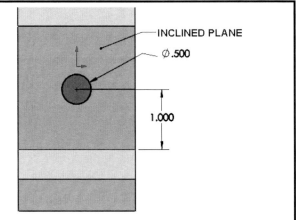

Figure 3-34. Creating the Circle on the Inclined Plane.

Click on the inclined face and return to a **Normal to** view of it, add a new **Sketch** and draw a larger **Circle** on it. **Dimension** it with a diameter of **0.80** inches. Next, click **Add Relations** and make this new circle **Concentric** with the edge of the through hole. Select the **Extruded Cut** command and use the **Blind** end condition **0.25** inches into the part. You now have a counterbore hole as seen in **Figure 3-35**.

You can now make the V-cut to add stress relief for the Toe Clamp. Select the flat horizontal face between the inclined faces. Change to a Top view. From the **Sketch tab** select the **Sketch** command; with the **Line** tool draw a small triangle shape approximately over the beveled edge of the Toe Clamp, as shown in **Figure 3-36**. Add the **Dimensions** indicated. If you apply them correctly, the sketch will turn *black*.

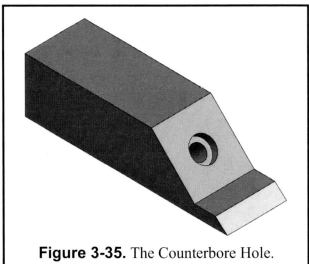

Figure 3-35. The Counterbore Hole.

Now you can make the V-Cut. Select the **Extruded Cut** command from the Features tab and use the **Through all** end condition. Make sure the cut goes into the part through the beveled end, and then click OK to finish. Use the **Rotate View** command to see it.

From the **FeatureManager Tree - Select** the **Right Plane then the Right View** and then select the **Sketch Command.**
Now Select the Centerline command and **Draw a Vertical Centerline** through the **Origin**.

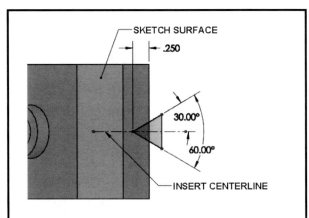

Figure 3-36. The Dimensions for the V-Cut.

Next, go to **Tools – Sketch Tools - Dynamic Mirror**. Your next step is to **Select** the **Rectangle** command and Sketch a Rectangle along the **Left Edge** of your model as shown in **Figure 3-37**. This rectangle is automatically mirrored on the Right Side because the Dynamic Mirror Command is active. From the Features Tab select **Extruded Cut – Through All** in both Directions, or use the **Through All-Both** in Direction 1. The resulting model will look like **Figure 3-39**.

The final feature to create is the large counter slot on the very top face of the Toe Clamp. Select the top face, add a new **Sketch** and draw the outer slot shape as shown in **Figure 3-40** using the **Straight Slot** sketch tool. **Add Relation** as follows:

The centers of the arcs have to be **Horizontal** with the **ORIGIN**.

Next, **Dimension** the slot by applying the values given in **Figure 3-40**. The sketch will be fully defined and turn *black*. Select the **Extruded Cut** command and make a **Blind** end condition **0.25** inches deep into the part. The 0.25 deep counter slot is created.

Select the same top face to start the final sketch. Activate the **Sketch** tab and select **Sketch**. Select the bottom face of the previous slot feature, and click on the **Offset** command. Use the Reverse direction option to make it inward by **0.20** inches to create a smaller slot than the previous shape. Select the **Extruded Cut** command and use the **Through all** end condition. Make sure the slot is cut in a downward direction, and then click OK to finish. (*Note:* This process for creating the counter slot was the reverse of the counterbore process, but they both worked correctly.) The Toe Clamp model is now finished as shown in **Figure 3-41** in an **Isometric** view.

Now you need to define the material for the **TOE CLAMP**. In the FeatureManager Tree, right click on **Edit Material**, expand the **Steel** materials category and assign **AISI Type A2 Tool Steel** to the **TOE CLAMP**.

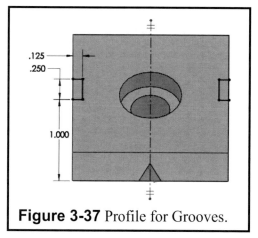

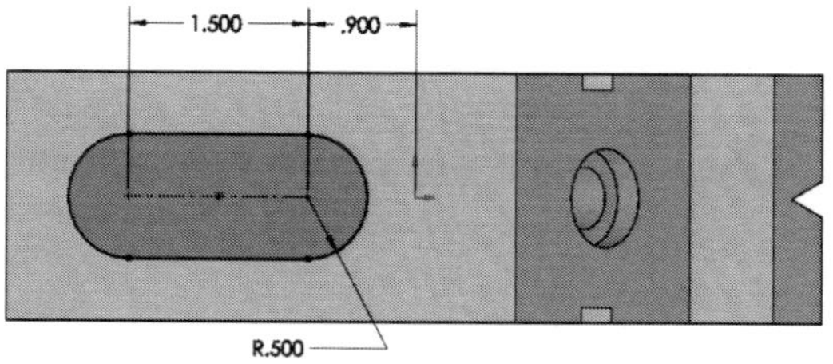

Figure 3-37 Profile for Grooves.

Figure 3-38 Grooves of the Toe Block.

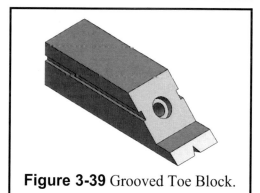

Figure 3-39 Grooved Toe Block.

Figure 3-40. The Counter Slot Feature.

Pull down **File,** select **Save As,** type in the part name **TOE CLAMP.sldprt**, select your proper folder, and then click **Save**. Add an isometric view in the **Title Block** drawing sheet like in previous labs. Save your drawing as **TOE CLAMP.slddrw**. **Print** a hard copy to submit to your lab instructor.

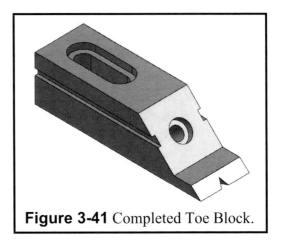

Figure 3-41 Completed Toe Block.

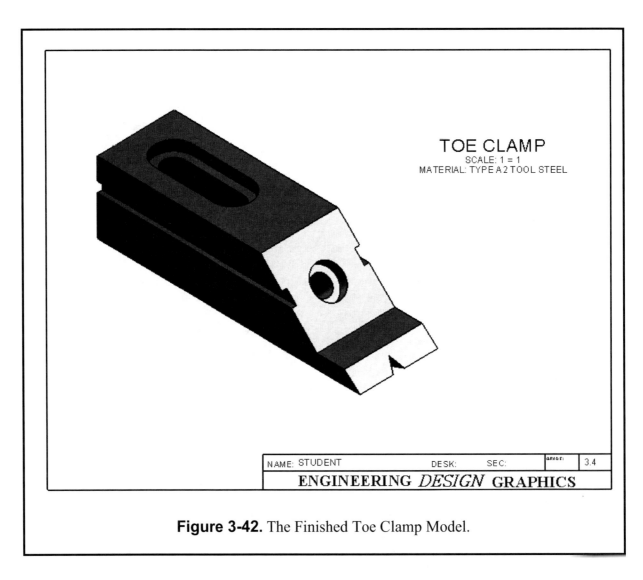

TOE CLAMP
SCALE: 1 = 1
MATERIAL: TYPE A 2 TOOL STEEL

NAME: STUDENT   DESK: SEC: GRADE: 3.4

ENGINEERING *DESIGN* GRAPHICS

Figure 3-42. The Finished Toe Clamp Model.

SUPPLEMENTARY EXERCISE 3-5: CONVEYOR RAMP GUIDE

Build a solid model of the figure below. Add an Isometric view to a Title Block and name it **"CONVEYOR RAMP GUIDE." Use the dimensions indicated below.**

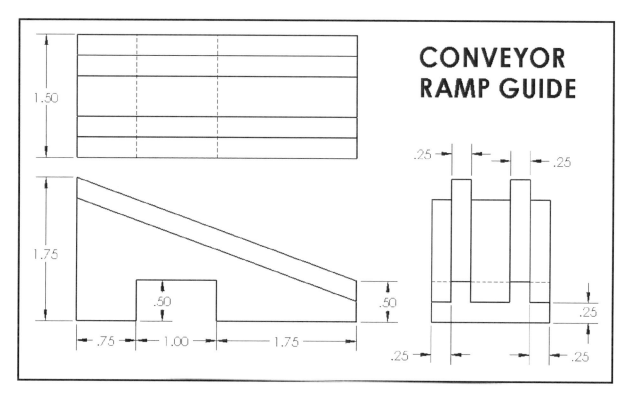

CONVEYOR RAMP GUIDE

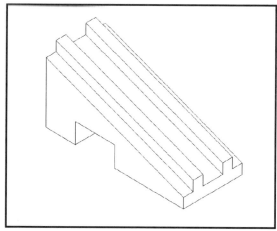

ISOMETRIC VIEW

SUPPLEMENTARY EXERCISE 3-6: DOUBLE SHAFT HANGER

Using the commands learned during the last two Units, build a 3D model of the figure below. Add an isometric view onto a Title Block drawing and title it **DOUBLE SHAFT HANGER**.

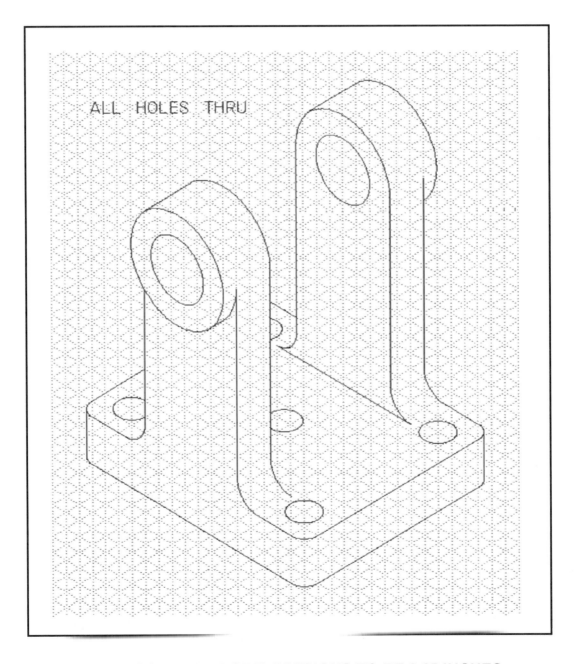

ALL HOLES THRU

ASSUME THE GRID DIVISIONS TO BE 0.25 INCHES.

INTRODUCTION TO LAB 4

In this lab you will be using many of the design features that make building solid models more efficient. Commands you will be using:

Draft - Creates a feature that tapers selected model faces by a specified angle, using either a neutral plane or a parting line. **NOTE:** You can also apply a draft angle as a part of an extruded base, boss, or cut feature.

Offset Plane - You can create planes in parts or assemblies. You can use planes to sketch, to create a section view of a model, for a neutral plane in a draft feature, and so on.

Offset - You can create sketch curves offset from one or more selected sketch entities, edges, loops, faces, curves, set of edges, or set of curves by a specified distance. The selected sketch entity can be changed to construction geometry. The offset entities can be bi-directional.

Convert Entities - You can create one or more curves in a sketch by projecting an edge, loop, face, curve, or external sketch contour, set of edges, or set of sketch curves onto the sketch plane. You can convert sketched entities into construction geometry to use in creating model geometry.

Face/Edge Fillet - Fillet/Round creates a rounded internal or external face on the part. You can fillet all edges of a face, selected sets of faces, selected edges, or edge loops.

Shell - The shell tool hollows out the part, leaves open the faces you select, and creates thin-walled features on the remaining faces.

Loft - Creates a feature by making transitions between profiles. A loft can be a base, boss, cut, or face.

Dome - Creates a loft type of feature that begins with the shape of the selected face and lofts to a zero feature at a specified height.

Sweep - Creates a base, boss, cut, or face by moving a profile (section) along a path, according to these rules:

- The profile must be closed for a base or boss sweep feature; the profile may be open or closed for a face sweep feature.
- The path may be open or closed.
- The path may be a set of sketched curves contained in one sketch, a curve, or a set of model edges.
- The start point of the path must lie on the plane of the profile.
- The section, the path, or the resulting solid cannot be self-intersecting.

Exercise 4.1: DRAWER TRAY

Go to your folder and open **ANSI-INCHES.prtdot**. A good practice is to immediately name your part and save the file in your folder. Pull down **File** again, select **Save As**, and then name it **DRAWER TRAY.sldprt**.

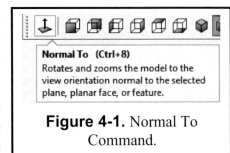

Figure 4-1. Normal To Command.

Select the **Top Plane** in which to work, then click on the **Normal To** command (**Figure 4-1**) which looks like a square plate with an arrow pointing upwards out of the center of the plate.

From the **Sketch Tab** select the **Sketch Command,** then draw a **Rectangle** at the origin and toward the upper right.

Using the Dimension command, constrain the rectangle to **4.5** inches high and **12** inches long. Now **Fillet** the four corners with a radius of **.375"**. The resulting sketch should look like the object in **Figure 4-2**.

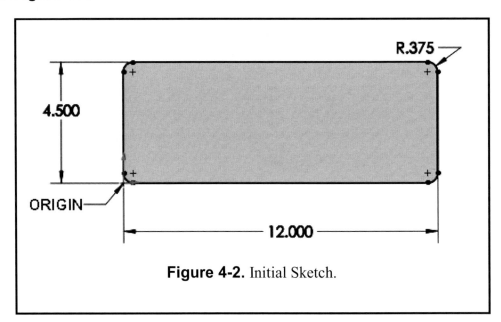

Figure 4-2. Initial Sketch.

From the Features tab select the **Extrude** command (**Figure 4-3**).

Do a **Blind Extrude downward** for **3** inches, with a **Draft Angle** of **5 Degrees** (See **Figure 4-4**). If the upward pointing arrow is highlighted, select it and pull it downward or click on the Reverse Direction option next to the "**Blind**" box. Click to finish. The part will look similar to **Figure 4-5**.

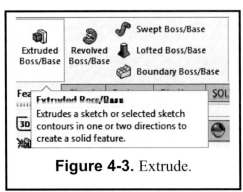

Figure 4-3. Extrude.

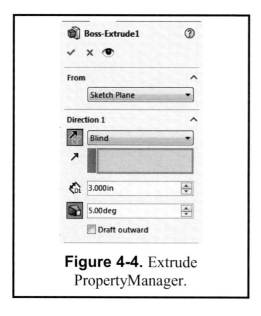

Figure 4-4. Extrude PropertyManager.

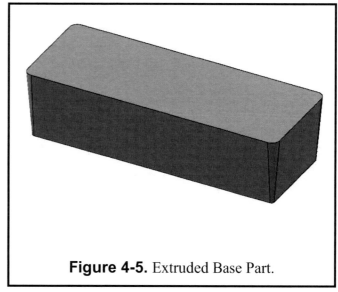

Figure 4-5. Extruded Base Part.

Slow double click on the "**Base Extrude**" feature in the FeatureManager and rename this feature as "**Tray Body**."

Now rotate the tray so you can see the bottom face. *To rotate the model, you can press down the middle mouse button and drag it in the screen.* Select the bottom face. Activate the **Feature Tab** and select the **Fillet** command (**Figure 4-11**). Enter a radius of **0.375** inches. All the edges of the selected face will be filleted.

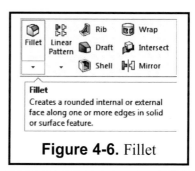

Fillet
Creates a rounded internal or external face along one or more edges in solid or surface feature.

Figure 4-6. Fillet

Click on the **Top** face of the Tray Body then click on the **Shell** command (**Figure 4-7**) from the **Features Tab** in the CommandManager. Set the shell thickness of **0.10** in the parameters box and click OK to finish. The model should now resemble the model in **Figure 4-8**.

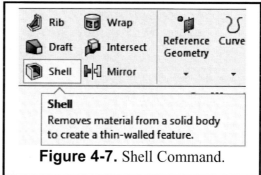

Shell
Removes material from a solid body to create a thin-walled feature.

Figure 4-7. Shell Command.

Drawer Tray Rim. In order to strengthen the upper edge of the drawer tray you will now create a rim around the top edge of the tray.

You might have to **Zoom** in to select the top thin face of the resulting tray (use the center wheel of your mouse). After you have selected the thin face, select the **Sketch command** from the **Sketch Tab**. Now select the **Convert Entities** command (**Figure 4-9**). This will project the outer edges of the selected face into new sketch as entities that can be used for a new feature.

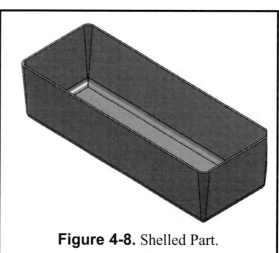

Figure 4-8. Shelled Part.

Next, select the **Offset Entities** command (see **Figure 4-10**) and select one of the projected entities; enter a distance of **0.10"**. If the yellow preview line is on the outside of the sketch, then activate the **Reverse** option so the

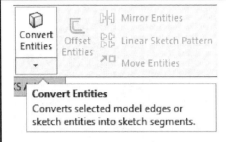

Figure 4-9. Convert Entities. Command.

Figure 4-10. Offset Command.

new offset geometry is inside the edge of the Drawer Tray. Extrude the sketch downward **0.625** inches. Make sure that the Merge Result box is checked. This creates a vertical rim around the upper edge of the tray.

Figure 4-11. Reference Plane Menu.

The next step is to place three equally spaced dividers inside the tray. Select the **Top Plane** in the FeatureManager. Then select **Insert, Reference Geometry, Plane**. Enter a distance of **0.25** and make sure the **Flip offset** checkbox is selected (**Figure 4-11**). Click **OK** to continue. This makes a new plane below the Top Plane that can be used to add a new sketch to draw the top view of the ribs. Rename this plane by slow double-clicking it and type **Dividers Plane**.

For the next step it might be convenient to change to a hidden line display mode. To do this go to the **Display Style** drop-down command and **Select** the **Hidden Lines Removed** (**Figure 4-12)**. Select the **Dividers Plane** in the FeatureManager and then the **Normal To** command to change to a top view orientation. Next, select the **Sketch** command and draw a **Rectangle across the inside edges** of the tray. Dimension the rectangle to **0.10** inches across and **3** inches from the outside edge of the Tray. See **Figure 4-13**. Select the **Linear Sketch Pattern** command, activate the "**Entities to Pattern**" box and select the four lines of the rectangle. Enter the following parameters: **Number (3); Spacing (3)**.

If the preview of the pattern is not inside the Tray, reverse the direction using the **Reverse Direction** command on the left of the **Direction 1** input box (**Figure 4-14**) to change the direction of the pattern. Click **OK** to continue.

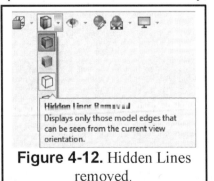

Figure 4-12. Hidden Lines removed.

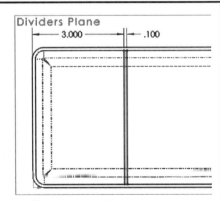

Figure 4-13. Draw and dimension a Rectangle.

You are now ready to extrude the patterns downward into the Tray. Select the **Features** tab, and **Select Extruded Boss/Base.** The extrusion arrow will be pointing upward. Click on the Reverse Direction option to make the extrusion go into the tray. Select the **Up To Next** end condition for **Direction 1** (**Figure 4-15**) and click **OK**.

Right click on **Material (not specified)** in the FeatureManager. Select **Edit Material**. Expand the **Plastics Tab** and select **PC High Viscosity**, then select **Apply** and **Close**. From the **Display Style** drop-down command **Select** the

Figure 4-14. Repeat Rectangle.

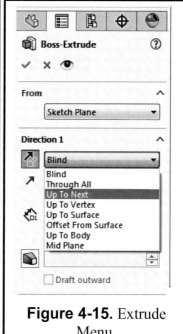

Figure 4-15. Extrude Menu.

Shaded with Edges style. Your completed model should look like **Figure 4-16**. Save your model to finish.

To finish this exercise, you should print a hard copy for submission to your instructor. Open your **TitleBlock-Inches.drwdot**. Save your drawing sheet as **DRAWER TRAY.slddrw**. Insert the isometric view into the **Title Block** drawing sheet like in previous labs. **Print** a hard copy to submit to your lab instructor. The drawing should look like **Figure 4-16**.

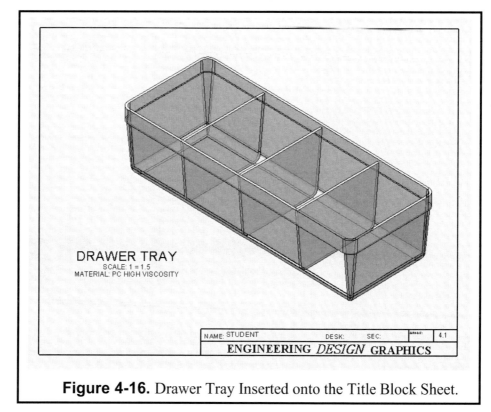

DRAWER TRAY
SCALE: 1 = 1.5
MATERIAL: PC HIGH VISCOSITY

| NAME: STUDENT | | DESK: | SEC: | DRGNO: | 4.1 |

ENGINEERING *DESIGN* GRAPHICS

Figure 4-16. Drawer Tray Inserted onto the Title Block Sheet.

Exercise 4.2: TAP-LIGHT DOME

Go to **File** and open **ANSI-INCHES.prtdot**. Pull down **File** again, select **Save As**, then name it **TAP-LIGHT DOME.sldprt** and save the file in your folder.

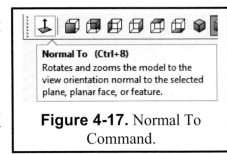

Normal To (Ctrl+8)
Rotates and zooms the model to the view orientation normal to the selected plane, planar face, or feature.

Figure 4-17. Normal To Command.

Select the **Top Plane**, and then change to a **Top View**. (See **Figure 4-17**.) Select the **Sketch** command and draw a **Circle** at the **ORIGIN** with a <u>**Diameter**</u> of **4.70** inches.

Select the **Extrude** command (**Figure 4-18**) from the Features tab. Do a **Blind Extrude** of **0.375** inches. Click on the **OK** button, your model will look similar to **Figure 4-19**.

Use a slow double click in "**Boss-Extrude1**" in the FeatureManager and rename this feature as "**Dome Base**."

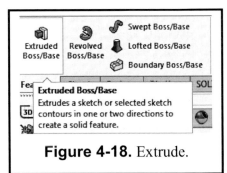

Extruded Boss/Base
Extrudes a sketch or selected sketch contours in one or two directions to create a solid feature.

Figure 4-18. Extrude.

Select the **Top** face of the **Dome Base**. Now go to the **Insert** tab **Select - Features** and **Select - Dome**. In the "**Dome**" PropertyManager window give the **Dome Height** of **1.20 inches** (See **Figure 4-20**) and click **OK**.

The dome operation should give you an object similar to **Figure 4-21**.

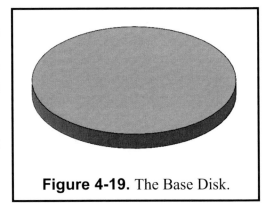

Figure 4-19. The Base Disk.

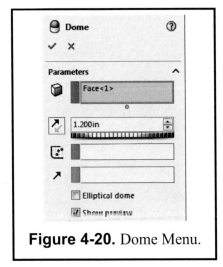

Figure 4-20. Dome Menu.

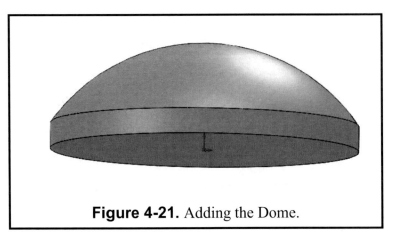

Figure 4-21. Adding the Dome.

Rotate the model so you can see the bottom face. Select the bottom face and select the **Shell** command from the Features tab (**Figure 4-22**). Set the shell thickness to **0.075"** in the parameters and click **OK** to finish.

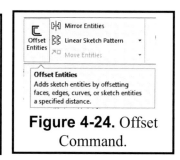

Figure 4-22. Shell Command.

You might have to zoom in to select the bottom thin face of the resulting shell. After you have selected the thin face, from the **Sketch Tab, Select** the **Sketch Command** and then select the **Convert Entities** command (**Figure 4-23**). This will project the outer edge of the face into a sketch entity that can be used for a new feature. Next, select the **Offset Entities** command (**Figure 4-24**) and select the new projected sketch entity; enter a distance of **0.20"** (**Figure 4-25**)

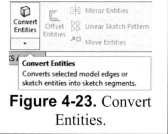

Figure 4-23. Convert Entities.

Figure 4-24. Offset Command.

outward so the offset sketch entity lies outside of the Dome (**Figure 4-26**). Make an **Extruded Boss** with this sketch upward a distance of **0.075"**.

This completes the basic shelled dome. The model will now look like **Figure 4-27**. The next step is to cut the four holes in the lip of the dome where fastening screws will pass to hold the body and the lens frame together. The holes in the lip of the dome are placed at uneven angles. This is done to ensure that the dome can only be placed on the body of the light fixture in one position.

Figure 4-25. Offset Menu.

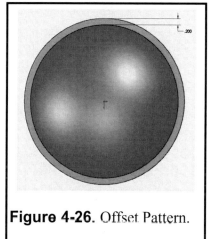

Figure 4-26. Offset Pattern.

Select the top face of the outer lip of the domed lens and change to a **Top view**. Select the **Sketch** command and draw a vertical **Centerline** through the origin. Next, **Draw four construction lines from the origin radially out, without capturing any automatic relations (Make sure the construction lines do not automatically snap to any relation). Next draw a circle with a diameter of 5.05"**, and with the circle selected

Figure 4-27. The Dome with a Lip.

check the **FOR CONSTRUCTION** option in the circle's PropertyManager as shown in **Figure 4-28**. The large bolt circle must be drawn since it is not the same diameter as the outer rim of the dome lip. Dimension the four radiating centerlines as shown in **Figure 4-29**.

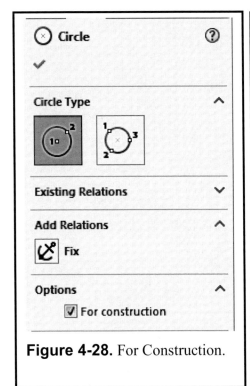

Figure 4-28. For Construction.

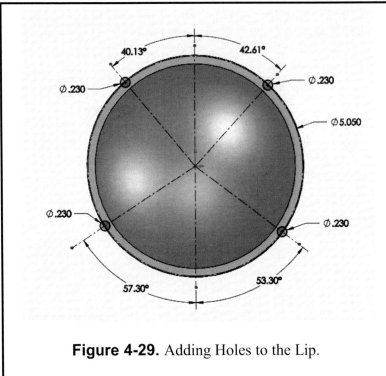

Figure 4-29. Adding Holes to the Lip.

Draw the four small holes at the intersections of the radial lines and the construction circle by using **Quick Snaps – Intersection Snap**. To get this menu, select the **Sketch – Circle** command, then **right click the mouse in an open area of the drawing screen and Select** the **Quick Snaps** option **(See Figure 4-30)**. Dimension the four circles and do the **Extruded Cut, Through All**.

Your final model should look similar to **Figure 4-31.**

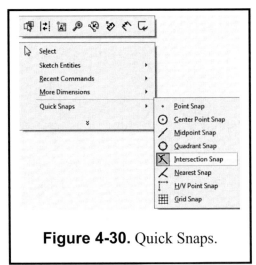

Figure 4-30. Quick Snaps.

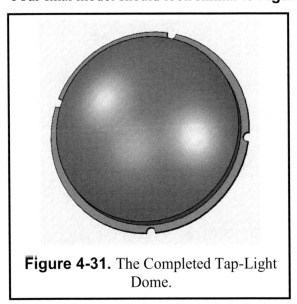

Figure 4-31. The Completed Tap-Light Dome.

Right click on **Material (not specified)** in the FeatureManager. Select **Edit Material**. Expand the **Plastics Tab** and select **ABS** and then select **Apply** and **Close**. Now save your model. Pull down the **File** menu and select **Save As**. On the "Save As" menu, select your appropriate file folder, type in the part name **TAP-LIGHT DOME.sldprt** and then click **Save**.

To finish this exercise, you should print a hard copy for submission to your instructor. First open your **TitleBlock-Inches.drwdot**. Follow the instructions given in Unit 1 for inserting the Isometric view onto a Title Block.

Save your drawing as **TAP-LIGHT DOME.slddrw**. Your final drawing should look similar to **Figure 4-32**.

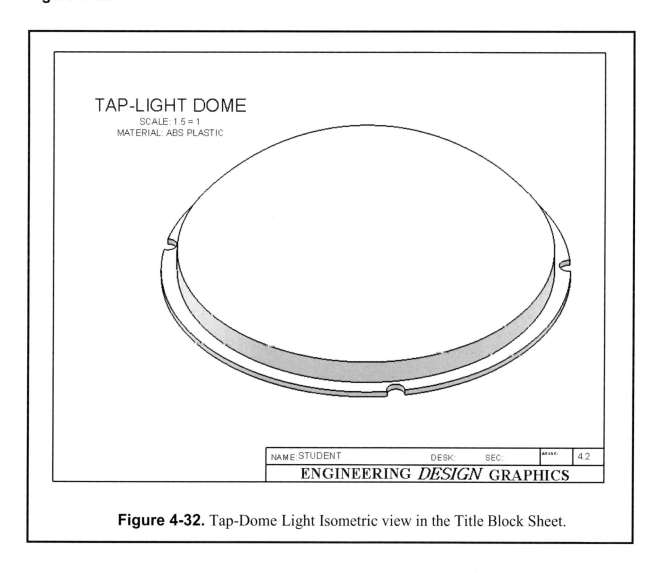

Figure 4-32. Tap-Dome Light Isometric view in the Title Block Sheet.

Exercise 4.3:
THREADS AND FASTENERS

THREAD TYPES

SQUARE THREADS & ACME THREADS These Threads have been used for many years to transmit power either from a turning motion or a linear motion. The uses have increased as more and more processes require this type of translation. Since the square thread is difficult to manufacture, the Acme thread has become the more commonly used thread for the purpose of translating power. **Figure 4-33** shows the profile of an Acme Thread and **Figure 4-34** shows a Square Thread profile.

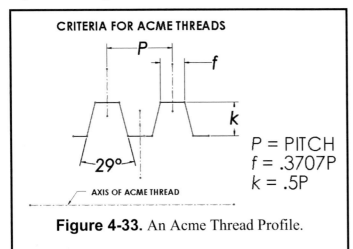

Figure 4-33. An Acme Thread Profile.

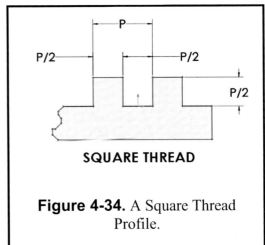

Figure 4-34. A Square Thread Profile.

BUTTRESS THREAD This thread form is a nonsymmetrical thread that is used where exceptionally high stresses lie along the axis of the threaded shaft. An example of this thread form is shown in **Figure 4-35.**

STANDARD V-THREADS - This thread form is the basis for most nuts and bolts; however, within this thread form there are many deviations that are used for special applications. Several of the most common standards are the American Standards and the Metric

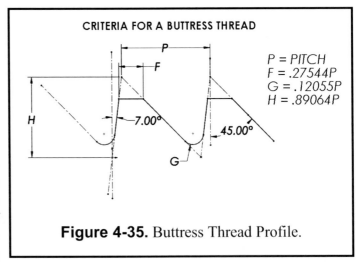

Figure 4-35. Buttress Thread Profile.

Standards. Within the American standards the most common series are "UNC" or Unified National Course (also known as "NC" National Course); "UNF" or Unified National Fine (also known as "NF" National Fine); and "UNEF" or Unified National Extra Fine (also known as "NEF" National Extra Fine). Within the Metric standards there is a coarse series and a fine series that are distinguished only by the different values of the pitch.

THREAD TERMINOLOGY

Following is a list of terms that are used extensively in association with threads:

➢ Major Diameter – the largest diameter of an internal or external thread.

➢ Pitch – The distance from one crest of a thread to the next crest.

➢ Pitch Diameter – the theoretical diameter at the point where the tooth width and gap width are equal.

➢ Minor Diameter - the smallest diameter of an internal or external thread.

➢ Threads per Inch – A number of threads in an inch that determines the pitch.

➢ Thread Depth – generally the difference between the major diameter and minor diameter divided by 2.

THREAD NOTES

Thread notes are the most critical portion of a thread representation. It gives all of the information that is necessary for a machinist to produce the thread or the manufacturer to select the correct fasteners for assembly. In **Figure 4-36** an American thread note is illustrated and in **Figure 4-37** is a metric thread note.

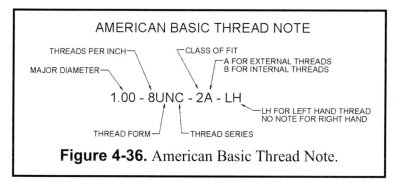

Figure 4-36. American Basic Thread Note.

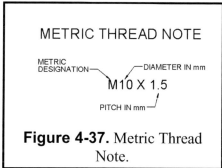

Figure 4-37. Metric Thread Note.

The dimensions of threads are based primarily on the number of threads per inch in the American system. The number of threads divided into one inch produces the pitch, which is also the basis in the metric system. In the exercise we will be using the criteria for internal and external threads. **Figure 4-38** shows the criteria required for external threads and **Figure 4-39** gives the specifications for the internal thread. The values for the different formulas are given in the various tables provided in Machinist or Engineering Handbooks. **Figure 4-40** is an example of such a table and includes the criteria that will be used in **Exercise 4.3**.

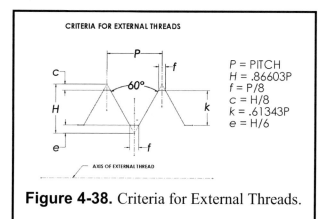

Figure 4-38. Criteria for External Threads.

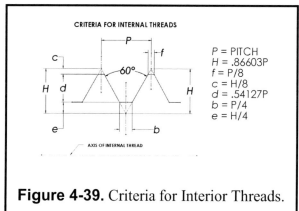

Figure 4-39. Criteria for Interior Threads.

UNIFIED COURSE THREAD SERIES PARTIAL TABLE					
Fractional Sizes	Basic Major Diameter	Threads per Inch	Pitch	Minor Diameter Internal Threads	Thread Depth
1/4	0.250	20	0.0500000	0.1959	0.0325
5/16	0.313	18	0.0555555	0.2524	0.0361
3/8	0.375	16	0.0625000	0.3073	0.0406
7/16	0.438	14	0.0714286	0.3602	0.0464
1/2	0.500	13	0.0769231	0.4167	0.5000
9/16	0.563	12	0.0833333	0.4723	0.0541
5/8	0.625	11	0.0909091	0.5266	0.0590
3/4	0.750	10	0.1000000	0.6417	0.0650
7/8	0.875	9	0.1111111	0.7547	0.0722
1	1.000	8	0.1250000	0.8647	0.0767
1 - 1/8	1.125	7	0.1428571	0.9704	0.0928
1 - 1/4	1.250	7	0.1428571	1.0954	0.0928
1 - 1/2	1.500	6	0.1666667	1.3196	0.1083

Figure 4-40. Unified Course Thread Table for Selected Sizes.

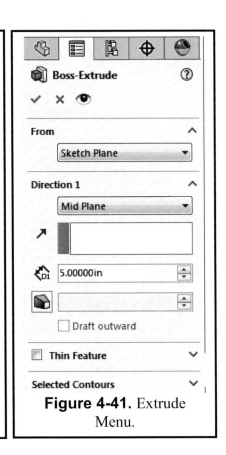

Figure 4-41. Extrude Menu.

Start a new part by going to your folder and opening **ANSI-INCHES.prtdot**. Pull down **File** again, select **Save As**, then name it **SHAFT THREAD.sldprt** and save the file in your folder. Go to **Tools – Options – Document Settings - Units** and change the units to **5 Decimal Places**.

Select the **Right Plane** and start a new **Sketch**. Draw a **Circle** at the **Origin** with a **Diameter** of **1.00** inch and **Extrude** it **5** inches using the **Mid Plane** end condition. The result will look like **Figure 4-42**.

Figure 4-42. The Resulting Shaft.

For the sake of proper clearances between interior and exterior threads, the exterior thread for a 1 inch thread is undersized by a tolerance of between 0.9966 and 0.9744. In our case we will use the mid-value of 0.9855. Change to a Left view orientation, select the round face on the left side of the shaft, add a new **Sketch** and **Draw** a circle with a diameter of **0.9855**. On the **Features** Tab, select the **Extruded Cut - Blind** option a distance of **3.25** inches. Activate the **Flip Side to Cut** option; see **Figure 4-43**.

Now we will cut threads on this shaft for **3.125** inches from the left end. **Select** the round face at the **Left End** of the shaft and from the **Features tab** select the **Chamfer** command. Using the **Thread Depth** formula in **Figure 4-38 on page 4-11** calculate the chamfer size: (.61343*.125) or **.0767"**.

Next, select the **Left** end of the shaft and select the menu **Insert - Reference Geometry - Plane** and enter the **Pitch** distance for a 1" thread from the Table in **Figure 4-40** (**0.125"**). The chamfer and the new plane should look like **Figure 4-44.**

Select **Plane 1**, or the plane you just created. Then activate the Sketch tab and Select **Sketch**. Select the large circular edge of the chamfer and select **Convert Entities**. Now go to the **Features** tab, click on the **Curves** command and select **Helix/Spiral**. In the PropertyManager options enter the following parameters and click **OK** to complete the helix.

> Defined By: **Height and Pitch**
> Parameters: **Constant Pitch**
>> Height: **3.25**
>> Pitch: **0.125**
> Check "**Reverse Direction**"
>> Start Angle: **270 degrees**
> Check: **Clockwise**

Since we started the helix at 270 degrees, the start is perpendicular to the Front plane, and therefore we can draw the tooth profile on the Front plane. In the FeatureManager Tree, select the **Front Plane**. Go to the Front View, go to the **Sketch Tab** and **Select Sketch**. Zoom into the left end of the shaft as shown in Figure 4-45. **Draw a centerline horizontally** along the bottom edge of the shaft. **Draw a vertical centerline** coincident with Plane1, and a **horizontal centerline** starting at the bottom of the thread. With the vertical centerline active, go to **Tools – Sketch Tools – Dynamic Mirror**. This adds two equal signs across the ends of the vertical centerline to indicate that the dynamic mirror function is active. See **Figure 4-45**.

Tooth Profile. **Sketch** half of the tooth profile on the right side of the centerline. The **Dynamic Mirror** command will automatically create the other half of the profile. **Dimension** your sketch to match the dimensions given in **Figure 4-46**. Make sure the wide base of the tooth profile extends slightly below the horizontal centerline. The dimension of the small end of the tooth

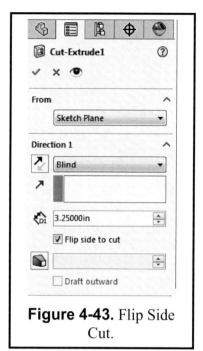

Figure 4-43. Flip Side Cut.

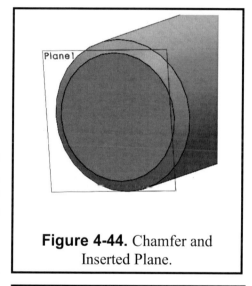

Figure 4-44. Chamfer and Inserted Plane.

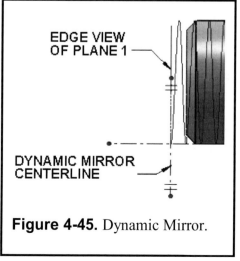

Figure 4-45. Dynamic Mirror.

profile is P/8. Since P = .125", .125/8 = **0.0156"**. The dimension from the horizontal centerline to the small end is 0.0767" obtained from the formula in **Figure 4-38** that shows that dimension is 0.61343P or 0.61343 * .125 = **.076678"**.

Click **Rebuild** to exit the sketch editing environment. Go to an isometric view to show how the **Swept Cut** operation will affect your threaded shaft. An isometric view should show the thread profile and sweep path similar to **Figure 4-47**.

Swept Cut. Now you are ready to create the completed thread. Activate the Features tab and select the **Swept Cut** command. In the PropertyManager select **Profile and Path**, using the mouse, first select the Tooth Profile and then the Helix curve. Click OK to complete the thread. Your part should look like **Figure 4-48**. Select the helix in the **FeatureManager** and click **Hide** from the context toolbar.

Now **Select** the round face at the **Right End** of the shaft and select the **Sketch Command,** then draw a **Hexagon** at the origin that is 1.5 inches across the flat edges of the hexagon. Draw the hexagon and locate a corner horizontal to the right of the origin. Next, use the smart dimension to define the distance from the top of the hexagon to the bottom to be **1.50** inches. This completes the sketch. Now **Extrude** the sketch away from the threaded end of the bolt to a depth of **39/64"** or **0.609375"**. This measurement is derived from the Standards for a 1" Thread. See **Figure 4-49** to view this operation.

Select the visible face of the Hexagon to begin your next sketch. Activate the Sketch Tab and select the Sketch Command. Draw a Circle at the origin and make it tangent to one of the flat edges of the Hexagon, **Figure 4-50. Select** the **Rebuild Command** and then **Select** the **Top Plane** then go to the top view to begin the next sketch. After you

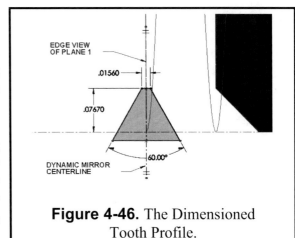

Figure 4-46. The Dimensioned Tooth Profile.

Figure 4-47. Helix Curve and Tooth Profile.

Figure 4-48. Completed Thread.

Figure 4-49. The Extruded Profile.

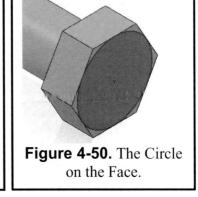

Figure 4-50. The Circle on the Face.

select the **Sketch Command, Select** the **Convert Entities Command. Select the previous sketch** to convert. After you click OK, **Select** the line that was projected and **select** "**For Construction**" in the line's **PropertyManager options**. All of this is required for this sketch. **Select** the **Line Command** and draw the Sketch at the upper point of the construction line. Draw a line at an angle across the corner of the screw's head then horizontally to the vertical face and then back down to the original point of the sketch. Dimension the sketch angle equal to 30 degrees as shown in **Figure 4-51**. Also draw a Center line horizontally through the origin. This completes the sketch. Select the Features Tab and **Revolve Cut** the 30 degree triangle about the centerline. The 1" bolt is now completed as shown in **Figure 4-52**.

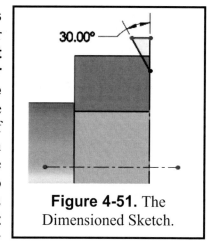

Figure 4-51. The Dimensioned Sketch.

CONSTRUCTION OF A CORRESPONDING NUT

There are many standards dictating the production of threads and fasteners. When beginning to construct fasteners it is advisable to consult official standards to verify accuracy. In the case of the 1" diameter nut that would fit the above threaded shaft, the hexagon of the nut is 1.5 times the major

Figure 4-52. The Completed Bolt

diameter of the shaft. Begin in the usual manner by going to your folder and opening **ANSI-INCHES.prtdot**. Pull down **File** again, select **Save As**, then name it **THREADED NUT.sldprt** and save the file in your folder. Go to **Tools – Options** and change the units to **5 Decimal Places**.

Now **Select** the **Right Plane** and select the **Sketch Command** then draw a **Hexagon** at the origin that is 1.5 inches across the flat edges of the hexagon. Draw the hexagon and locate a corner horizontal to the right of the origin. Next, use the smart dimension to define the distance from the top of the hexagon to the bottom to be **1.50** inches. Now referring to **Figure 4-40, Page 4-12** you will draw a **Circle** at the origin to be equal to the Minor Diameter for Internal Threads which is **0.8647** inches. This completes the sketch. Now **Extrude** the sketch at **Mid-Plane** to a depth of **55/64"** or **0.859375"**. This measurement is derived from the Standards for a 1" Thread. Your model should look like **Figure 4-54**.

Select the visible face of the Hexagon to begin your next sketch. Activate the Sketch Tab and select the Sketch Command. Draw a Circle at the origin and make it tangent to one of the flat edges of the Hexagon, **Figure 4-55**.

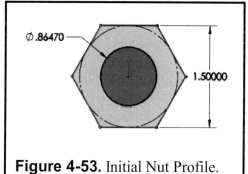

Figure 4-53. Initial Nut Profile.

Figure 4-54. The Extruded Profile.

Select the **Rebuild Command** to exit the sketch, then **Select** the **Top Plane**, and change to a top view to begin the next sketch. After you select the **Sketch Command**, **Select** the **Convert Entities Command**. **Select the previous sketch** to project and click OK to finish. **Select** the line that was generated and **select "For Construction"** in the **PropertyManager**. **Select** the **Line Command** and begin the Sketch at the upper endpoint of the construction line. Draw a line at an angle across the corner of the nut then horizontally to the vertical face of the nut and then back down to the original point of the sketch. Dimension the sketch angle to equal to 30 degrees as shown in **Figure 4-56**. Also draw a Center line horizontally through the origin. This completes the sketch. Select the Features Tab and **Revolve Cut** the sketch about the centerline. If needed, select the centerline going through the origin and click OK. Now **Select** the **Mirror Feature** and select the **Right Plane**. This will make a mirror copy of the Revolved Cut on the other side of the nut. **See Figure 4-57**.

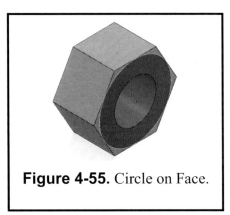

Figure 4-55. Circle on Face.

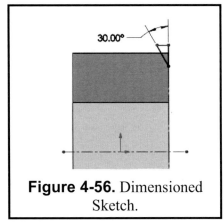

Figure 4-56. Dimensioned Sketch.

The next operation is to construct a **Chamfer** inside both ends of the hole, equal to the Thread Depth which is in the Chart found in **Figure 4-45**, **Page 4-13** for the 1 inch Diameter, (**0.0767**). Select the inside face of the nut, enter the chamfer dimension and click OK to finish.

Select the face on the Right side of the nut. Then go to **Insert – Reference Geometry – Plane**. Place it 1 Pitch distance away from the selected face. Pitch distance is **0.125"** and click OK. Select Plane1, click the **Sketch Command** and **Convert** the small diameter. Now select **Insert – Curve – Helix/Spiral**. In the **Defined By: Select Height and Pitch**. For the Height enter something a little larger than 1" and the **Pitch** is 8 per inch or **.125"**. The **Start Angle** should be at **270 Degrees**. Select OK to

Figure 4-57. Revolved Cut.

accept these parameters, then **Select** the **Front Plane** and go to the front view. **Select** the **Sketch Command** and draw **Two Centerlines**. The first one should be horizontal and coincident at the bottom of the Hole. Change to a **Hidden Lines Visible** display mode to see where to place this line. The second centerline is vertical along the edge view of **Plane1**. While this line is selected, go to **Tools – Sketch Tools – Dynamic Mirror**. This puts an equal sign on both ends of the vertical. Draw half of the tooth profile starting on the centerline and the other side will be automatically duplicated across the centerline. Dimension the tooth profile as shown in **Figure 4-59**. These dimensions are derived from the formulas in **Figure 4-38**. Rebuild your solid model to exit the sketch mode. Go to the **Features Tab** and **Select Swept Cut**. First select the tooth profile, then the Helix to cut the thread into the nut. Your final part should look like **Figure 4-60**. Select the helix in the **FeatureManager** and click **Hide** from the context toolbar.

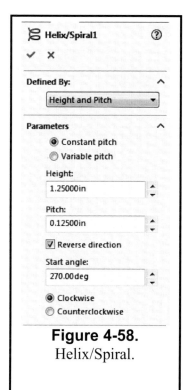

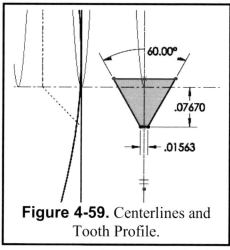

Figure 4-59. Centerlines and Tooth Profile.

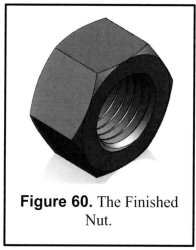

Figure 60. The Finished Nut.

Figure 4-58. Helix/Spiral.

Save your part as **Threaded Nut.SLDPRT**.

To finish this exercise, you should print a hard copy for submission to your instructor. First open your **TitleBlock-Inches.drwdot**. Insert the two isometric views into the **Title Block** drawing sheet as in **Figure 4-61**. Save your drawing as **SCREW THREADS.slddrw**. **Print** a hard copy to submit to your lab instructor.

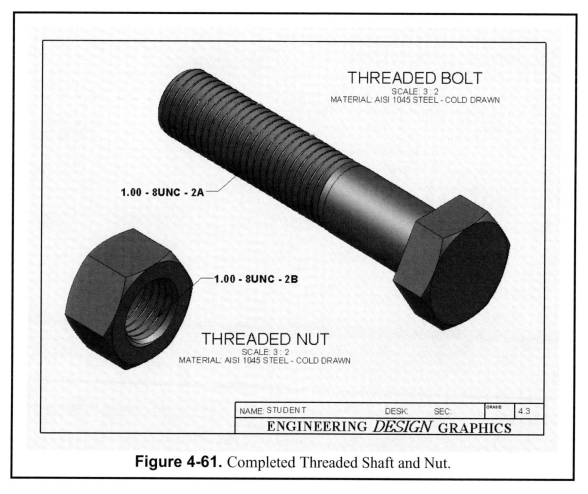

Figure 4-61. Completed Threaded Shaft and Nut.

Exercise 4.4: JACK STAND

Go to **File** and open **ANSI-INCHES.prtdot**. Pull down **File** again, select **Save As**, then name it **JACK STAND.sldprt** and save the file in your folder.

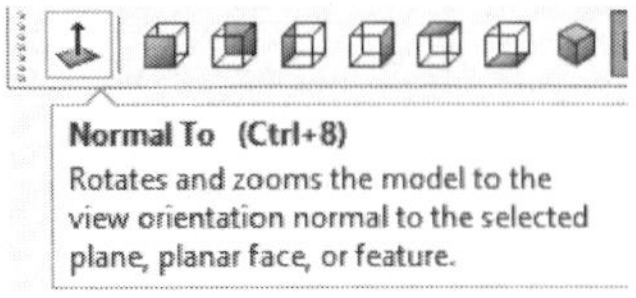

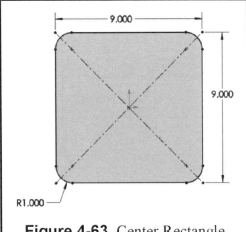

Figure 4-62. Normal To Command.

Select the **Top Plane,** change to a **Top View** or click on the **Normal To** command shown in (**Figure 4-62**).

Select the **Sketch** command and from the **Rectangle** drop-down menu, select the **Center Rectangle** command, draw the rectangle starting at the origin and dimension it as shown in **Figure 4-63**. Add **1"** Fillets to the corners.
Select the **Rebuild** command (the stop light) to exit the sketch mode.

In the next step you will have to create a new reference plane. To better see what is going to happen, change to an isometric view. Select the **Top Plane** and then select **Insert, Reference Geometry, Plane**. Enter the distance of **6.5"** and click OK to continue.

After you have created the reference plane, select **Plane1** in the **FeatureManager Tree**, and then the **Sketch** command. Change to a **Top view**. In the new sketch draw a **Circle** that is **4.0"** in diameter, centered on the origin.

Figure 4-63. Center Rectangle Sketch.

Select the rebuild command (the stop light) to exit the sketch mode. **Select** the **Top Plane** again and then select **Insert, Reference Geometry - Plane** again. Enter a distance of **10"**. This plane will be labeled **Plane2**. Change to a **Top** view, select **Plane2** and **Sketch** a **Circle** on it that is **3.0"** in diameter, centered on the origin. Go to an isometric view to see the resulting sketch profiles as shown in **Figure 4-64**.

Select the rebuild command (the stop light). **Select** the **Lofted Boss/Base** command in Features menu. (See **Figure 4-65**.)

To better visualize the **Loft** operation next, change to an **Isometric** view.

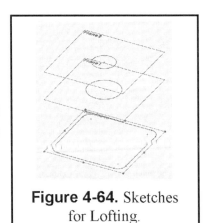

Figure 4-64. Sketches for Lofting.

Figure 4-65. Loft Command.

Select the **Lofted Boss/Base** command from the **Features** tab, and then select the three sketches in order from bottom to top (or top to bottom, but in order). The order of the sketch selection will appear in the **Loft** PropertyManager (See **Figure 4-66**). As you select the sketches a preview of the loft will be shown on the screen. If it appears as the figure in **Figure 4-67**, click OK to complete the loft.

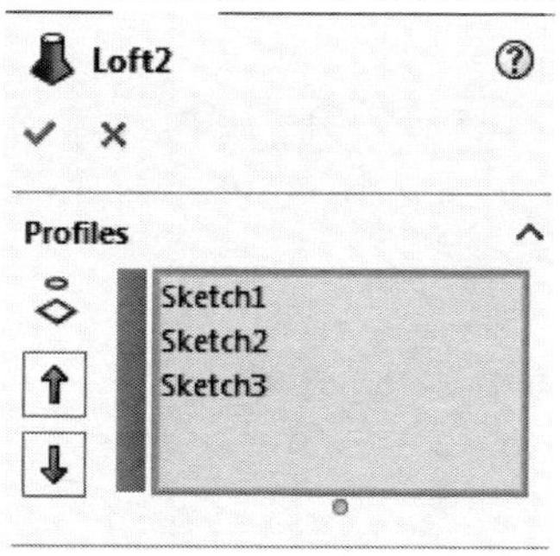

Rotate the model so you can see the **bottom face. Select** the **bottom face** and click on the **Shell** command (**Figure 4-68**) in the **Features** tab. Make the shell **0.1875"** thick and click OK to finish.

Figure 4-66. Loft PropertyManager.

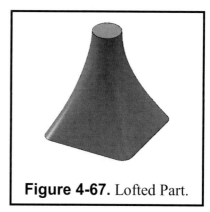

Figure 4-67. Lofted Part.

Several other operations must be performed in order to make this a viable design. The top must be reinforced for a threaded adjustment screw and the sides are to be hollowed out to lighten the weight of the jack stand. (The thread formation has been omitted from this design).

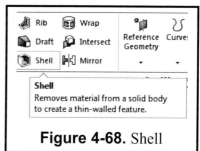

Figure 4-68. Shell

Select the **Top Face** of the jack stand and **Select** the **Normal-To** view orientation command. Select the Sketch tab and Select **Sketch,** then use **Convert Entities** and make a **Boss Extrude** downward for **3"**. **Select** the **Top Face** again and **Sketch** a **1.25** diameter **Circle**. Make an **Extruded Cut, Through all** feature to create the hole as shown in **Figure 4-69**.

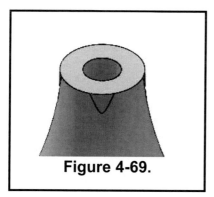

Figure 4-69.

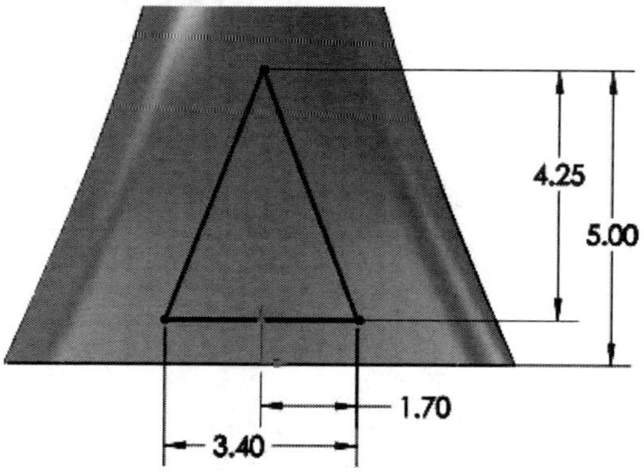

The last two operations are the openings that need to be cut in the sides of the jack stand. **Select** the **Front Plane** and **Select**

Figure 4-70. Sketching the Triangle.

Normal-To. Sketch the **Triangle** and set the dimension constraints as shown in **Figure 4-70**. Also add a **vertical relation** between the top point of the triangle and the origin. **Make** a **Cut Extrude, Through All** in **Both Directions.**

Since you will be using the exact same sketch in the Right Plane, select the **Previous Sketch** in the **FeatureManager,** go to the **Windows Edit** menu and select **Copy** (see **Figure 4-71**). Next, go to the **FeatureManager Tree**, select the **Right Plane**, and from the **Windows Edit** menu select **Paste**. Select the **Sketch** that resulted from this operation and repeat the **Cut Extrude, Through All** in **Both Directions** to complete the model. Your final model should look similar to **Figure 4-72**.

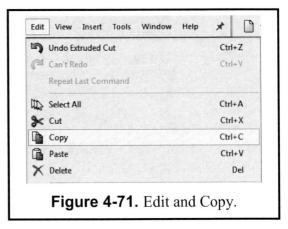

Figure 4-71. Edit and Copy.

Now save your model. From the **File** menu select **Save As**. On the "Save As" menu, select your appropriate file folder, type in the part name **JACK STAND.sldprt**, then click **Save**.

To finish this exercise, you should print a hard copy for submission to your instructor. First open your **TitleBlock-Inches.drwdot**. Follow the instructions given in Unit 1 for inserting the isometric view onto a Title Block. Save your drawing as **JACK STAND.slddrw** in your designated folder.

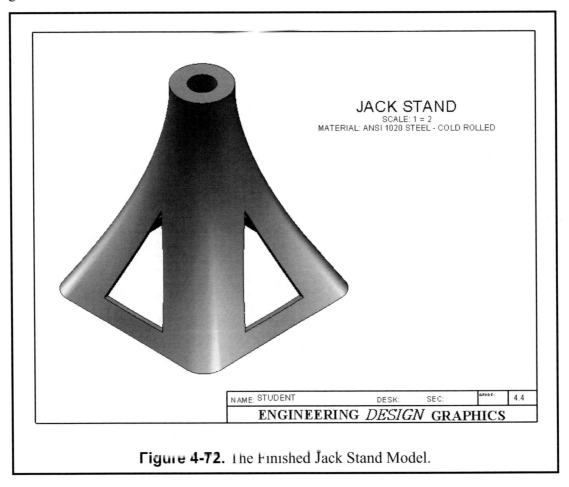

Figure 4-72. The Finished Jack Stand Model.

SUPPLEMENTARY EXERCISE 4-5: TAPE DISPENSER

Using the **ANSI-INCHES.prtdot**, draw the profile of the Tape Dispenser and extrude it 0.75 inches. Select the front face and fillet it to 0.05 inches. Select the back face and the two interior faces as indicated in the second figure and SHELL the solid to a thickness of 0.10 inches. Open your **TitleBlock-Inches.drwdot**. Follow the instructions given in Unit 1 for inserting the isometric view onto a Title Block. Save your drawing as **TAPE DISPENSER.slddrw** in your designated folder.

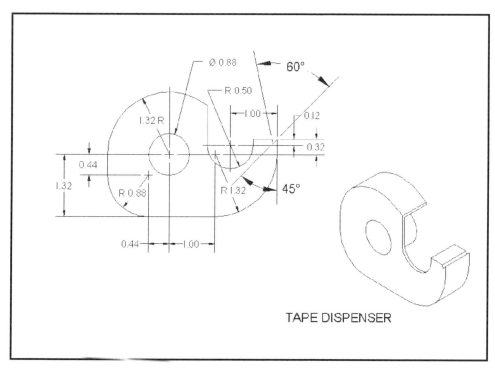

TAPE DISPENSER

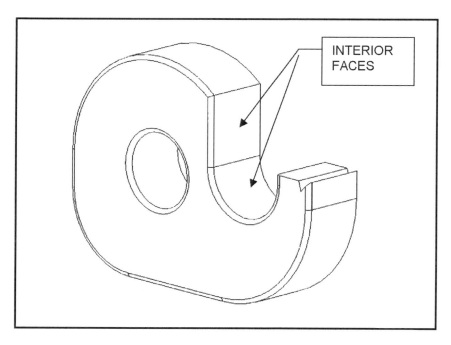

INTERIOR FACES

SUPPLEMENTARY EXERCISE 4-6: FUNNEL

Build a model of the Funnel below. Begin in the **Top Plane**. To get the desired model, each section must be lofted separately. Activate the right plane and sketch a guide curve connecting the left hand edges of the first two sketches. Repeat the process for the other two lofts. When the lofting process is completed, select the top and the bottom faces and shell the model to a wall thickness of 0.10 inches. Insert it onto a Title Block and title it **FUNNEL.slddrw**.

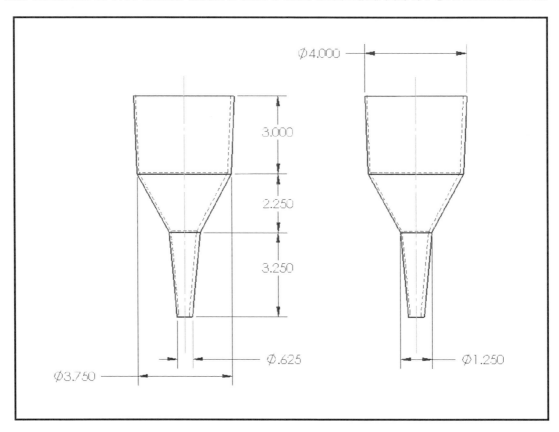

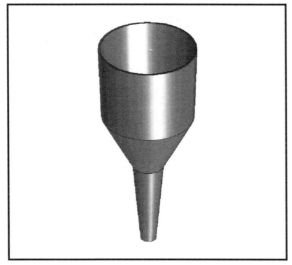

SUPPLEMENTARY EXERCISE 4-7:
TWO INCH THREAD – UNC

1. Make a 2" diameter shaft 5 inches long.
2. The major screw thread diameter is from 1.9868 – 1.9424. Cut the shaft down to 1.9646 for three inches.
3. Screw thread note - 2-4 ACME-2A X 3.
4. Add a chamfer equal to the Thread Depth (the "k" value in the Definition of Terms) on the end of the shaft that will be threaded.
5. Add a reference plane one pitch "P" distance away from the end of the shaft.
6. Select the reference plane and convert the 1.9646 diameter circle.
7. Insert – curve – helix spiral with the parameters – height and pitch. The height will be 3.25 and the pitch will be .25, with the starting angle being at 270 degrees – clockwise.
8. Select the Front plane and draw a centerline along the lower edge of the shaft and a vertical centerline through the edge of the inserted plane. Select the vertical Centerline and use the Dynamic Mirror function.
9. Sketch the tooth profile according to the given parameters.
10. Use the Features – Swept Cut function to cut the thread.

ACME THREAD SPECIFICATIONS					
Nomial Size	Basic Major Diameter	Threads per Inch	Pitch	Thread Depth d = .5P + 0.01	Width of Space at Bottom of Thread W = .3707P - .0052
2	1.9646	4	0.25	0.135	0.087475

Values from Thread Table

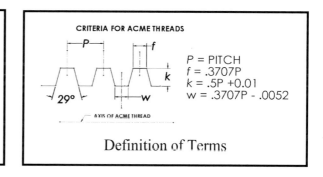

CRITERIA FOR ACME THREADS

P = PITCH
f = .3707P
k = .5P +0.01
w = .3707P - .0052

Definition of Terms

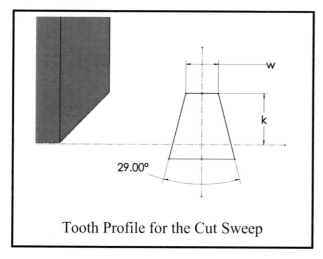

Tooth Profile for the Cut Sweep

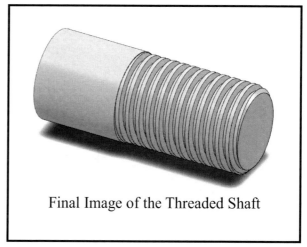

Final Image of the Threaded Shaft

NOTES:

Computer Graphics Lab 5: Assembly Modeling and Mating

In Computer Graphics Lab 5, you will build several parts of an assembly. You will then assemble the parts and fit them properly together using mating relations available in SOLIDWORKS. This introductory section will get you started and then two exercises, the Terminal Support Assembly and the Swivel Eye Block Assembly, are provided.

ASSEMBLY FILE

In the previous exercises, you started with a **File, New, Part** command sequence to start a new part file. After making solid model parts in SOLIDWORKS, we can make an assembly with multiple different components. You make a new assembly selecting the Assembly template from the File, New menu. Also, you can make a 2D detail drawing of parts and assemblies selecting the Drawing template as shown in **Figure 5-1**. This is the usual way to start an Assembly.

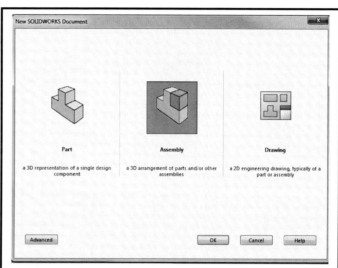

Figure 5-1. Starting a New Assembly File.

TILING THE WINDOWS

There are different ways to start the assembly. One way is to open all the parts to the assembly in SOLIDWORKS, in addition to having the new assembly file open. You can then **Tile Vertically** or **Horizontally** the windows (see **Figure 5-2**) and then simply "Drag and Drop" the parts into the assembly file. The first part placed into the assembly is fixed and cannot be moved. The others can be fixed in relation to the first part by using the Mate commands.

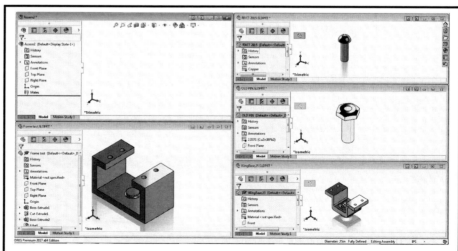

Figure 5-2. Tiling the Windows Vertically to Begin the Assembly.

THE ASSEMBLY TOOLBAR

When you are working in the assembly mode, the CommandManager includes tabs with assembly tools. It contains the following commands:

Insert Components – Adds an existing part or sub-assembly to the assembly.

Mate – Positions two components relative to one another.

Linear Component Pattern – Patterns components in one or two linear directions.

Smart Fasteners – Add features to the assembly using the SOLIDWORKS Toolbox Library of standard hardware.

Move Component – Moves a component within the degrees of freedom defined by its mates.

Show Hidden Components

Assembly Features – Creates various assembly features.

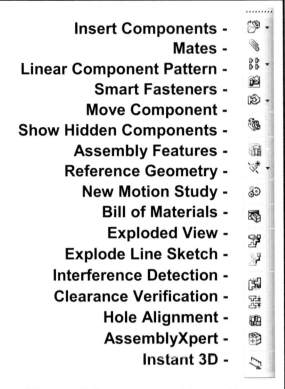

Insert Components -
Mates -
Linear Component Pattern -
Smart Fasteners -
Move Component -
Show Hidden Components -
Assembly Features -
Reference Geometry -
New Motion Study -
Bill of Materials -
Exploded View -
Explode Line Sketch -
Interference Detection -
Clearance Verification -
Hole Alignment -
AssemblyXpert -
Instant 3D -

Figure 5-3. The Assembly Toolbar.

Reference Geometry – Reference geometry commands: Plane, Axis, Coordinate System, Point, and Mate Reference.

New Motion Study – Inserts a new motion study.

Bill of Materials

Exploded View – This separates the components into an exploded view.

Explode Line Sketch – This can be used to add or edit a 3-D line sketch, showing the relationship between exploded components.

Interference Detection – This can be used to detect any interference between components in an assembly.

Clearance Verification – Verifies clearance between components.

Hole Alignment – This checks assembly hole alignment.

AssemblyXpert – Displays statistics and checks the health of the current assembly.

Instant 3D - This enables dragging of handles, dimensions, and sketches to dynamically modify features.

MATE TYPES

You can mate parts and sub-assemblies by selecting the **Mate** command in an assembly. Based on the entities selected to be mated, the valid mate options will be presented. Valid mate references include model faces, edges, planes, axes, sketch entities, etc.

The valid mating relations are:

Coincident - positions selected faces, edges, and planes flat, touching or aligned with other faces, edges, planes, vertices, etc. Coincident positions two vertices so they share a common point.

Perpendicular - places the selected items at a 90-degree angle to each other.

Tangent - places the selected items in a tangent mate (at least one selection must be a cylindrical, conical, or spherical face).

Concentric - places the selections so that they share the same axis.

Parallel - places the selected items so they lie in the same direction and remain a constant distance apart from each other.

Distance - places the selected items parallel at a specified distance between them.

Angle - places the selected items at the specified angle to each other.

Symmetry - places the selected items at an equal distance from a plane of symmetry.

VIEWING ASSEMBLIES

When you are building assemblies with many components, sometimes viewing specific components or assembly details becomes difficult. One way to improve component visibility is to select an assembly component, and from the pop-up context toolbar you can select the Change Transparency command, to make the selected component transparent, as seen in **Figure 5-4**.

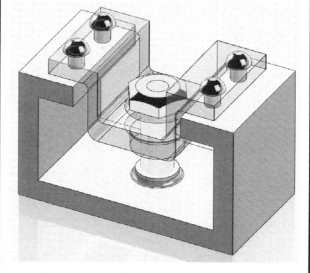

Figure 5-4. Changing a Component's Transparency in an Assembly.

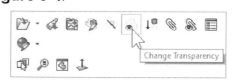

EXPLODING ASSEMBLIES

Assembly components can be exploded using the **Exploded View** command from the assembly toolbar. In the **Explode**

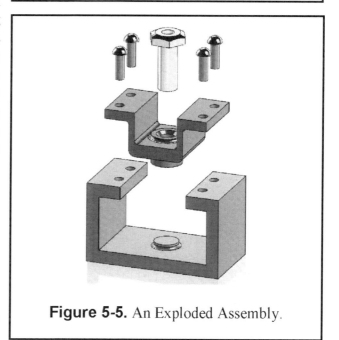

Figure 5-5. An Exploded Assembly.

PropertyManager you can define the parameters for each assembly explode step, as shown in **Figure 5-5**. More on exploding assemblies will be covered in **Unit 8**.

Exercise 5.1: TERMINAL SUPPORT ASSEMBLY

The Terminal Support Assembly has three major components: the Frame, the Wing Base and the Pin. It also contains four standard rivet parts for attachment to the frame. You will start by designing the Frame part. After that you will model the Wing Base, the Pin and then finally create one version of the rivet. Finally, you will mate all of them in an assembly, including adding the rivet four times.

FRAME

Go to your folder and open **ANSI-INCHES.prtdot**. A good practice is to immediately name your part and save the file in your folder. Pull down **File** again, select **Save As**, and then name it **FRAME.sldprt**. Next, select the **Front** plane in the FeatureManager tree and also change to a **Front** view orientation. Add a new **Sketch** and draw a vertical **Centerline** through the origin. Then go to the menu **Tools, Sketch Tools**, and select **Dynamic Mirror**. Now, draw the right half of the image in **Figure 5-6**. The left side of the sketch will be automatically added. When the sketch is finished, extrude it **4.00 inches** using the **Mid Plane** end condition.

Change to a Top view, select the **TOP** recessed face in the Frame and add a new Sketch. Draw the four holes in the sketch according to the dimensions given in **Figure 5-7**. Make an **Extruded Cut** using the **Up to Next** end condition; it will cut the part only to the next face.

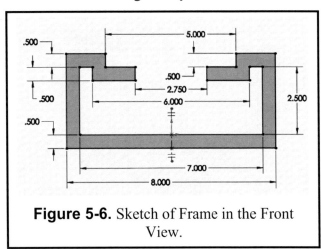

Figure 5-6. Sketch of Frame in the Front View.

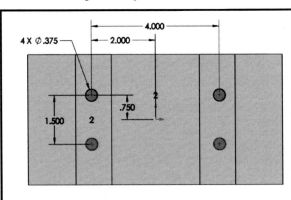

Figure 5-7. Sketch of Holes in the Top View.

Select the **Top** face of the **Base** and change to a **Top** view orientation. Add a new **Sketch** and Draw a Circle **1.50 inch** in Diameter starting at the origin. Next, **Extrude** this circle **.25 inches upward**. Add a **0.125 inch Fillet** at the intersection between the cylinder and the base of the frame. Add a Chamfer to the two top inner edges of the frame and Fillets to the inner edges of the base. These are all .25". See **Figure 5-8** for a view of the completed Frame.
In the FeatureManager Tree, right click on **Edit Material** and assign **AISI 4340 Steel, annealed** to the Frame.

You can now save the completed part. Pull down **File**, select **Save As**, and save the part as **FRAME.sldprt** in your folder. Now **Close** your Frame part. You are now ready to start the next part for the assembly.

WING BASE

Go to your folder and open **ANSI-INCHES.prtdot**. Immediately pull down **File** again, select **Save As**, and then name it **WING BASE.sldprt**. Select the **Front** plane in the FeatureManager tree and add a new sketch in it. If needed, change to a **Front** view orientation.

Draw the sketch indicated in **Figure 5-9** as indicated using the **Line** tool.

Use the sketch **Fillet** to round the four corners with the radii (**0.125** or **0.500**) as indicated. **Select** all the geometry simultaneously including the centerline going through the origin, and **Mirror** the profile about the centerline to complete the sketch.

Select the **Extrude Boss/Base** command and make an extrusion using the **Mid Plane** end condition with **2.500** inches distance. Click OK to continue. Your part will look like **Figure 5-11**. Rename the "Base-Extrude" feature as **Base** in the FeatureManager.

Select the top face of the "Base" feature and change to a **Top** view orientation. Add a new **Sketch** in this face and draw a **Circle** on it, centered at the origin, with a <u>diameter</u> of **1.250"**. Select the **Extrude Boss/Base** command. Set "Direction 1" (up) to **Blind** with a distance of **0.25** inch. Set "Direction 2" (down) to **Blind** with a distance of **1.00** inch. Click OK to finish. Rename this feature **Big Boss**. Select the top face of this big boss, add a new **Sketch** and draw a **Circle** on this face, centered at the origin, with a <u>diameter</u> of **0.75"**.

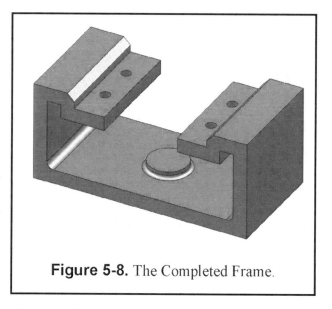

Figure 5-8. The Completed Frame.

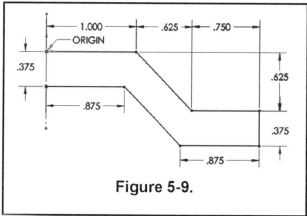

Figure 5-9.

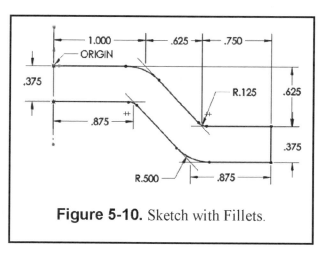

Figure 5-10. Sketch with Fillets.

Select the **Extrude Cut** command and cut the hole using the **Through All** end condition. Rename this new hole feature **Big Hole**. You now have a model that looks like **Figure 5-11** in a **Rotated View** orientation.

You now need to add a chamfer and fillets to some of the edges in the model. Review **Figure 5-12** to determine which edges to chamfer and fillet. To better view these edges, change to a **Hidden Lines Visible** view. Select the **Chamfer** command from features toolbar, select the edge to chamfer, and set the following parameters on the "Chamfer" Property-Manager:

Select "Angle Distance"
Distance = **0.10** inches
Angle = **45** degrees
Then click OK to finish.

Next, select the **Fillet** command from the features toolbar, and select the two edges where the Big Boss feature intersects the Base. Set the fillet radius to **0.125** inches. Then click OK to finish. You can change back to a **Shaded with Edges, Trimetric** view to better see the chamfer and fillets as shown in **Figure 5-13**.

The final features to add to the Wing Base are the four holes on the two wings where the rivets will be attached. Change to a Top view orientation, select the top face of the left wing feature and start a new **Sketch**. Draw a **Circle** on the face as indicated in **Figure 5-14**. Go to **Tools – Options – Document Properties** and unselect the Grid Display. Use the **Dimension** tools to precisely locate this first hole.

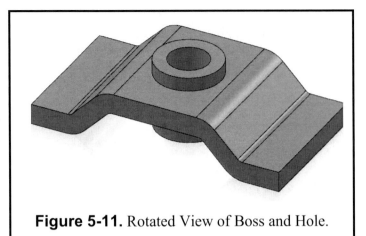

Figure 5-11. Rotated View of Boss and Hole.

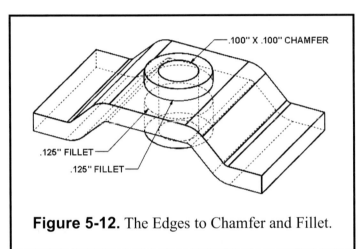

Figure 5-12. The Edges to Chamfer and Fillet.

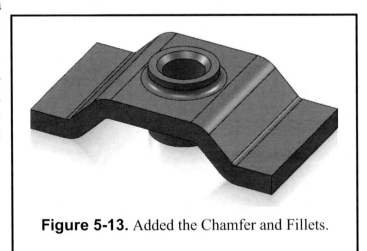

Figure 5-13. Added the Chamfer and Fillets.

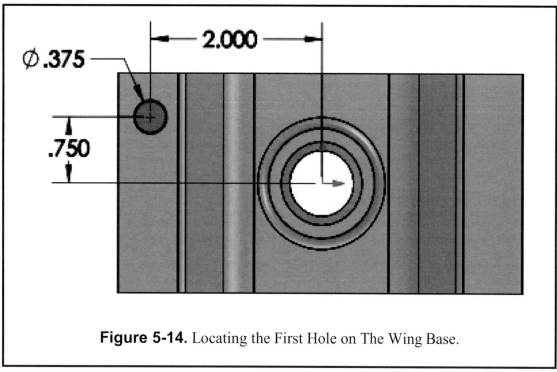

Figure 5-14. Locating the First Hole on The Wing Base.

Now select **Tools**, **Sketch Tools**, and **Linear Sketch Pattern**. In the menu, supply the following parameters:

> **Direction 1**
> > Number = **2**   Spacing = **4.000** inches   Angle = go to the **right**
>
> **Direction 2**
> > Number = **2**   Spacing = **1. 500** inches   Angle = go **down**

If the preview looks like **Figure 5-15** click **OK** to finish the linear pattern.

Select the **Extrude Cut** command and cut the holes using the **Through All** end condition. Rename this new feature **Four Holes**. You now have a finished model of the Wing Base that looks like **Figure 5-16** in a **Trimetric** view orientation.

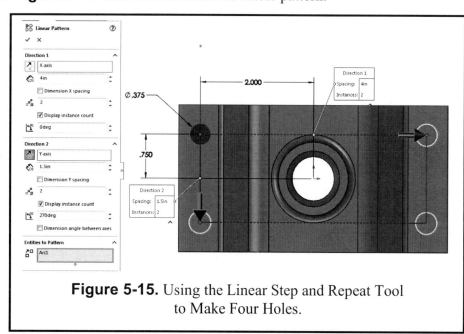

Figure 5-15. Using the Linear Step and Repeat Tool to Make Four Holes.

Now we need to define the material for the Wing Base. In the FeatureManager Tree, right click on **Edit Material** and assign **Chrome Stainless Steel** to the Wing Base.

You can now save the completed Wing Base. Pull down **File**, select **Save As**, and save the part as **WING BASE.sldprt** in your folder. Now **Close** your Wing Base part. You are now ready to start the next part for the assembly.

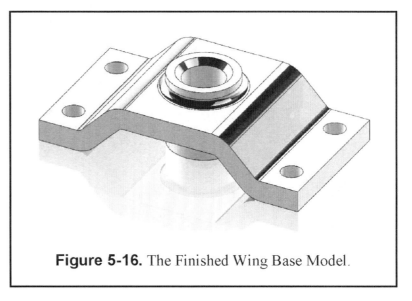

Figure 5-16. The Finished Wing Base Model.

PIN

Go to your folder and open **ANSI-INCHES.prtdot**. Immediately pull down **File** again, select **Save As**, and then name it **PIN.sldprt**. Next, select the **Front** plane in the FeatureManager tree and change to a **Front** view orientation. With the Front plane selected, click on the **Sketch** command to make a new sketch.

Draw the sketch profile as shown in **Figure 5-17** and use the **Line** tool to draw the profile. Add a vertical **Centerline** through the **ORIGIN**. Add the sketch **Fillet** to the corner with a radius of **0.125** as indicated. Select the **Revolve Boss/Base** command and revolve the sketch profile **360** degrees. Rename this feature as **Pin Base** in the FeatureManager.

You now need to add the hex head feature to the top of the Pin. Change to a **Top** view, select the top flat face of the Pin and add a new **sketch**. From the **Sketch** tab select the **Polygon** tool. Make sure the number of sides is set to 6. Center the hexagon at the origin and drag a vertex horizontally out to the right and make it coincident with the edge at the perimeter of the Pin, as shown in **Figure 5-18**.

Select an **Isometric** view orientation to better see the next operation. Select the **Extruded Cut** command, and set the end condition to **Through All**. Also select the **Flip side to cut** option to cut outside of the sketch profile. Click OK to complete the Cut Extrude. The finished part will look like **Figure 5-19**. Rename this feature to **Hex Cut** in the FeatureManager.

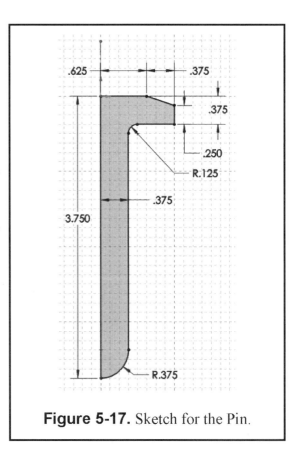

Figure 5-17. Sketch for the Pin.

In the FeatureManager Tree, right click on **Edit Material** and expand **SOLIDWORKS DIN Materials**. Then expand **DIN Copper Alloys** and assign **2.1020 (CuSn6)** to the Pin. The Pin is now complete as can be seen in **Figure 20.**

You can now save the final part. Pull down **File**, select **Save As**, and save the part as **PIN.sldprt** in your folder. Now you can **Close** the part.

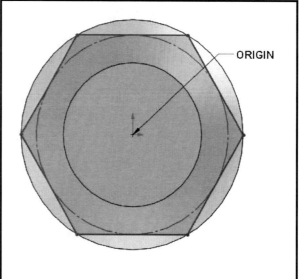

Figure 5-18. Adding a Hexagon to Cut the Hex Head.

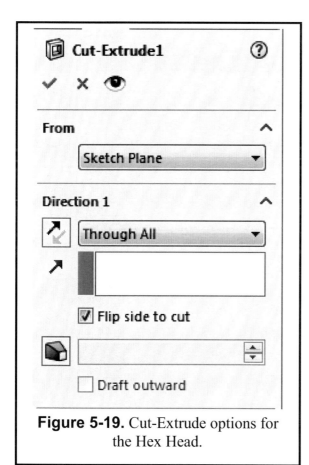

Figure 5-19. Cut-Extrude options for the Hex Head.

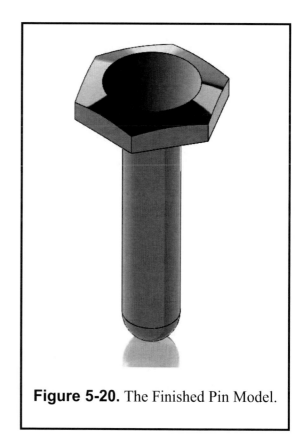

Figure 5-20. The Finished Pin Model.

5-9

RIVET

The Rivet is a simple part and you can create it in a number of ways. The method chosen here is to simply sketch a half-profile and then revolve it around a centerline. There are different types of rivets (flat head, pan head, etc.). For this case, you will design a "Button Head" rivet.

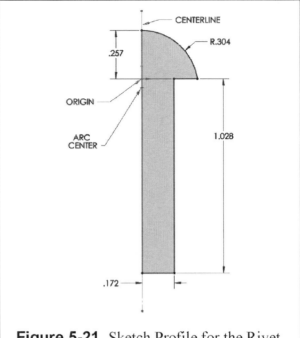

Study the geometry in **Figure 5-21**. Go to your folder and open **ANSI-INCHES.prtdot**. A good practice is to immediately name your part. Pull down **File** again, select **Save As**, and then name it **RIVET.sldprt**. Select the **Front** plane icon in the FeatureManager tree and change to a **Front** view orientation. You will draw and dimension the sketch profile.

Add a new **Sketch** in the **Front** plane and use the **Line** and **Centerpoint Arc** tools to draw the profile indicated in **Figure 5-21**.

Figure 5-21. Sketch Profile for the Rivet

Remember to draw a vertical **Centerline** through the **ORIGIN**. Then use the **Dimension** tool to set all the given dimensions. If they are placed correctly, all lines should turn *black* meaning the geometry is completely defined. ***Note:*** **The center of the arc is below the origin.** If it happens to be at the origin in your case, then remove the "Coincident" relation between them by clicking the **Display/Delete Relations** command from the Sketch tab.

Select the **Revolve Boss/Base** command and revolve the sketch profile **360** degrees. Rename this feature as **Rivet Base** in the FeatureManager. The finished rivet is shown in **Figure 5-22** in an Isometric View.

In the FeatureManager Tree, right click on **Edit Material** and assign **2.0090 (Cu-DHP)** from the DIN Copper Alloys Materials list to the Rivet.

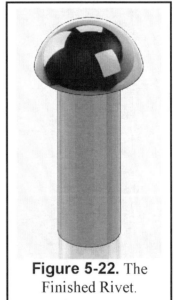

You can now save the completed Rivet. Pull down **File**, select **Save As**, save the part as **RIVET.sldprt** in your folder and close the part. You are now ready to start the assembly.

Figure 5-22. The Finished Rivet.

TERMINAL SUPPORT ASSEMBLY

Pull down **File**, **New** and select **Assembly**. Pull down **File** again, select **Save As**, and then save it as **TERMINAL SUPPORT**. Pull down **File** and **Open,** one-by-one, the four parts of the assembly: **Frame, Wing Base**, **Pin**, and **Rivet**. You now have five files open in SOLIDWORKS. Pull down **Window** and select **Tile Vertically**. You can now see all four files on the screen as shown in **Figure 5-23**.

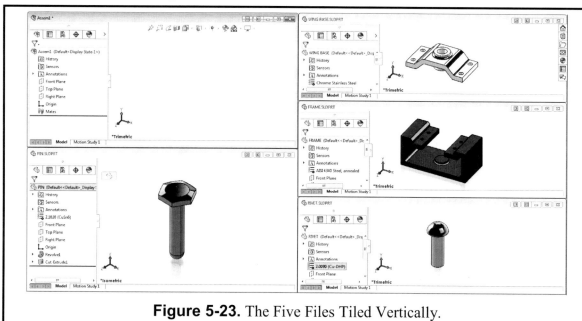

Figure 5-23. The Five Files Tiled Vertically.

Starting with the **Frame** file, **Select** the part's name in the FeatureManager and "Drag and Drop" it **into the assembly's FeatureManager**. The reason is to make the first part's principal planes aligned with the Assembly's principal planes. **NOTE**: The first part dropped into the assembly is **FIXED** and cannot be moved unless its constraints are deleted. It will be the base to which all other parts are mated. Now "Drag and Drop" the **Wing Base in the graphics area**, approximately as seen in **Figure 5-24**, then add the **PIN**. Finally, "Drag and Drop" the **RIVET <u>four times</u>** into the assembly. You can try to line them up when you "Drop" them into the assembly, but more importantly you will mate them next. You no longer need the part files open so you can **Close** them. Then maximize the assembly file window. Your assembly should now look similar to **Figure 5-24**.

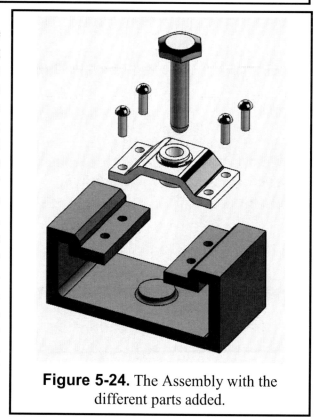

Figure 5-24. The Assembly with the different parts added.

You will mate the Frame and the Wing Base first. Select the **Assembly Tab** and click on the **Mate** command (it looks like a paper clip), and the "Mate" PropertyManager is displayed as indicated in **Figure 5-25**. Now select a cylindrical hole on the Frame and the corresponding hole in the Wing base. Do not select the edges of the holes. Click on the **Concentric** mate and click OK. Repeat this process for one more set of corresponding holes. This will constrain the Wing Base with the Frame along the axes of the holes. The next operation will establish constraints in relation to the top face of the Frame and the bottom face of the upper wing of the Wing Base. Still in the **Mate** command, select the top face of the Frame. Rotate the assembly so you can see the bottom face of the Wing Base and select the corresponding face, and select the **Coincident** mate.

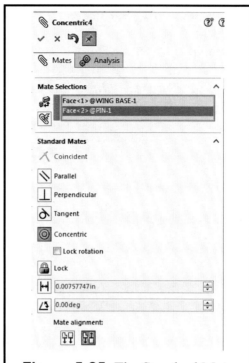

Figure 5-25. The Standard Mate Options.

Now select the outer cylindrical face of the Pin (the face that will go into the hole) and then select inner cylindrical face of the big hole on the Wing Base. Now select the **Concentric** mate and click OK. The Pin is now concentric with the hole and you should see it move over a little. You can look at it from a **Top** view if you wish to check it.

Repeat this **Concentric** mate four times, one-by-one, between each Rivet and its respective small hole on the wing faces. Use an **Isometric** view and **Zoom** in as needed to complete these mates.

Now try something else. **Cancel** the **Mate** command and click-and-drag the **Pin** or one of the **Rivets** in the screen. You will see that its movement is constrained in a vertical direction that corresponds to the axis of its respective hole.

You will now mate the parts to the Wing Base. Select the **Mate** command again. Select the flat face on the lip of the big hole. Next, you need to select the matching flat face at the bottom of the

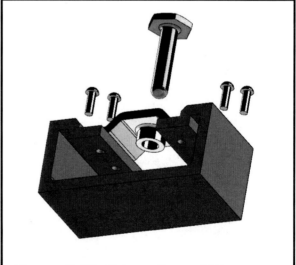

Figure 5-26. Using a Rotated View to see the Bottom Face of the Hex Head.

hex head, but it may be difficult to select. Use the middle mouse button or the arrow keys to **Rotate** the assembly view until you can see the face, and select the bottom of the hex head as shown in **Figure 5-26**. Select the **Coincident** mate and click OK. You should see the Pin move into the correct position on the **Wing Base** and its vertical motion has been restricted.

Cancel the Mate command, and return to an **Isometric** view orientation. Use the **Move Component** command from the **Assembly** tab, or simply click-and-drag to move it. Notice the pin cannot move up or down, but it can rotate about its axis, since that degree-of-freedom is not constrained. Now repeat this **Coincident** mating process for the four rivets, one-by-one. In each case, mate the flat face in the bottom of the **Rivet** head with the top **flat wing** face of its respective hole. **Rotate the View** as needed to make the selections. After you have completed these four mates, try to **Move** the components until you are satisfied that the assembly has been mated correctly. You are now finished adding mates and your Terminal Support Assembly is complete.

Now pull down **File**, select **Save As**, and save the assembly as **TERMINAL SUPPORT.sldasm** in your folder. Insert a Shaded Isometric view of the Terminal Support Assembly into your **Title Block** drawing sheet.

You will now create a Bill of Materials and add Balloons with Part names attached to the assembly. To begin, go to Tools – Options – Document Properties and make the changes that are shown in **Figure 5-27**. Set Leader style and Frame style to **0.0098in**. The leader display for Single/Stacked Balloons and Auto Balloons should both be Bent. Under the Single balloon selection box, set the style to Circular and the size to 2 Characters. Now click on **Font**. And set the Font to **Arial – Regular** – and the **Height to 0.25in**. Once all of these settings are made, click the **OK**.

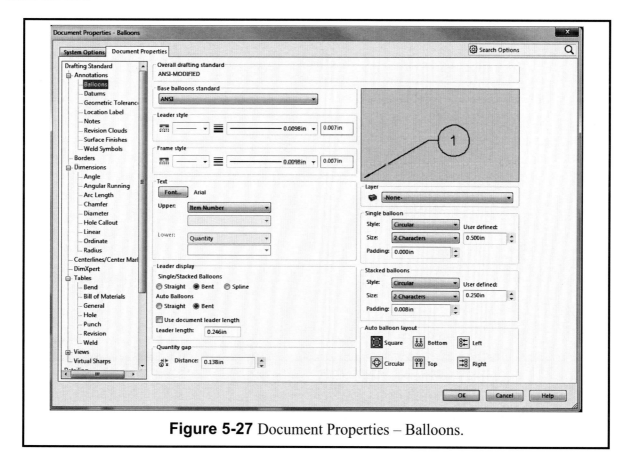

Figure 5-27 Document Properties – Balloons.

Select **Insert–Annotations – Balloon** and carefully select the Frame, the Wing Base, the Pin and one Rivet. Circles with the respective Part numbers will appear. After all of the Balloons are placed, select **Insert, Annotations, Note** (without a Leader) and place the part name (18 pt) next to the corresponding balloon.

Now select **Insert – Tables – Bill of Materials.** Once you select the assembly, the Bill of Materials PropertyManager is displayed. Select options Parts only, display all configurations of the same part as one item; for the border, the first number should be changed to 0.0098 and the second number is unchanged. Click OK to accept the parameters and locate the Bill of Materials on the drawing sheet. Place it in the upper left corner of your drawing

Now save your drawing as **TERMINAL SUPPORT.slddrw** and **Print** this sheet (see **Figure 5-28**). Now close your session, pull down File and click close. You are finished.

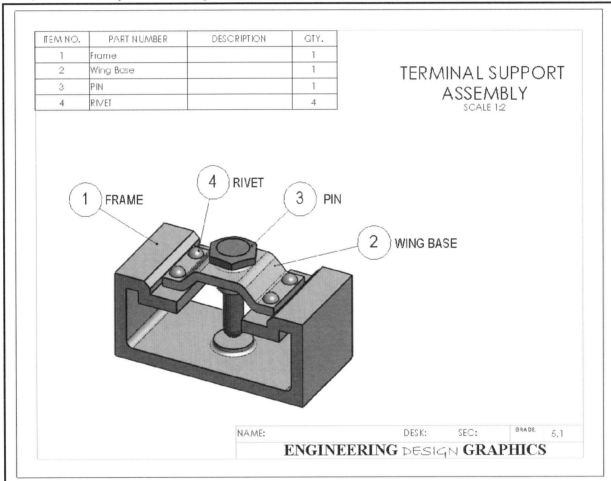

ITEM NO.	PART NUMBER	DESCRIPTION	QTY.
1	Frame		1
2	Wing Base		1
3	PIN		1
4	RIVET		4

TERMINAL SUPPORT
ASSEMBLY
SCALE 1:2

Figure 5-28. The Finished Terminal Support Assembly on the Title Sheet.

Be sure to save your assembly files for use in later lab exercises such as "Kinematics Simulation" where you will need to explode an assembly and make an animation file

Exercise 5.2: PULLEY ASSEMBLY

The Pulley Assembly is a common design used to hoist objects with a mechanical leverage advantage. It has four major components: an Eye Hook, a Pulley Sheave, a Spacer, and a Base Plate that is used twice. It is assembled using small rivets that are peened on one end to secure them and hold the components together. You will start the design by designing the Base Plate, which controls many of the dimensions of the other components.

BASE PLATE

Study the geometry of the Base Plate in **Figure 5-29**. Go to your folder and open **ANSI-INCHES.prtdot**. Immediately pull down **File** again, select **Save As**, and then name it **BASE PLATE.sldprt**. Next, select the **Front** plane in the FeatureManager tree and change to a **Front** view orientation. Your sketch will be symmetrical about a vertical centerline.

Add a new **Sketch** in the **Front** plane and use the **Line** and **Circle** sketch tools to draw the profile shown in **Figure 5-29**. Start with a **Circle** at the origin with a radius of **2.125** inches. Then draw three **Lines**, approximating the tangent points. Click the **Add Relation** icon and add a **Tangent** relation between the bottom circle and the left line, then add a **Tangent** relation between the bottom circle and right line. Use **Trim** to get rid of part of the circle not needed. Draw the two small **Circles** on the top of the profile, draw the center **Circle**, and make sure the top line is **Horizontal**. Finally, use the **Dimension** tool to completely define the sketch. If done correctly, the lines should turn *black*. Select **Extrude Boss/Base,** use a **Blind** end condition and distance of **0.1250** inches (one-eighth inch). In the FeatureManager Tree, right click on **Edit Material** and assign **AISI 4340 STEEL, NORMALIZED** from the SOLIDWORKS Materials – Steel list. The Base Plate is finished as shown in **Figure 5-30**.

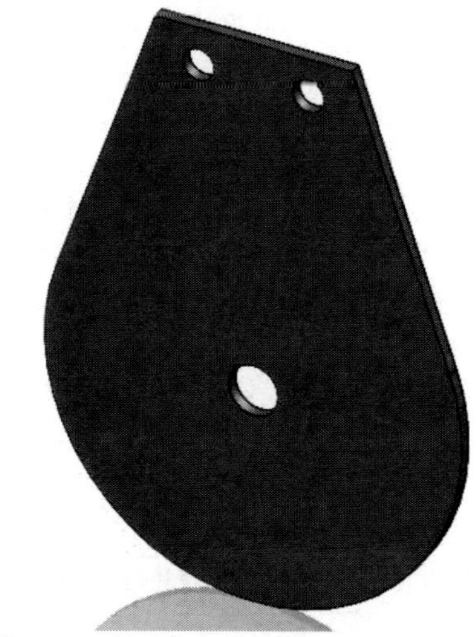

Save your **BASE PLATE.sldprt** and **Close** your file.

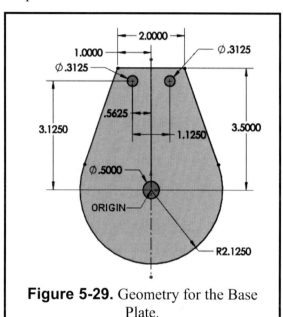

Figure 5-29. Geometry for the Base Plate.

Figure 5-30. Finished Base Plate.

PULLEY

Study the geometry of the Pulley in **Figure 5-31**. Go to your folder and open **ANSI-INCHES.prtdot**. Immediately pull down **File** again, select **Save As**, and then name it **PULLEY.sldprt**. Select the **Right Plane** in the FeatureManager tree and add a new sketch. The view orientation will be automatically changed to a **Right** view.

Use the **Line** and **Centerpoint Arc** sketch tools to draw the profile shown in **Figure 5-31**. Draw a **Vertical and Horizontal Centerline** through the **Origin**. The profile is symmetrical about the vertical centerline so you can use the **Dynamic Mirror** command found under **Tools – Sketch Tools**.

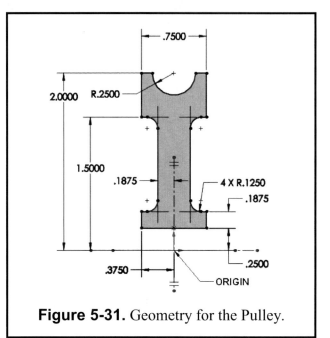

Figure 5-31. Geometry for the Pulley.

Use the **Dimension** tool to define all the geometry in place. If done correctly, the lines should turn *black*. Now select the **Revolve Boss/Base** command and make the revolved feature **360** degrees about the horizontal centerline. You now have the pulley part of the assembly.

In the FeatureManager Tree, right click on **Edit Material** and expand **SOLIDWORKS DIN Materials**. Then expand **DIN Copper Alloys** and assign **2.0375 (CuZn36Pb3)**. Then click the **Apply and Close buttons**. The Pulley is finished as shown in **Figure 5-32**. Be sure to **Save** your **PULLEY.sldprt** before you **Close** your file.

Figure 5-32. The Finished Pulley.

SPACER

Study the geometry of the Spacer in **Figure 5-33**. Go to your folder and open **ANSI-INCHES.prtdot**. Immediately pull down **File** again, select **Save As**, and then name it **SPACER.sldprt**. Next, select the **Front Plane** in the FeatureManager tree and also click the **Front** view orientation.

Add a new **Sketch** in the **Front** plane and use the **Center Rectangle** and **Circle** sketch tools to draw the profile shown in **Figure 5-33**. Then use the **Dimension** tool to define the sketch geometry. Be sure to dimension the position of the Spacer relative to the <u>origin</u>. If done correctly, the lines should turn *black*.

Select **Extrude Boss/Base,** use a **Mid Plane** end condition and distance of **1.000** inches. *Note*: A Mid Plane end condition will extrude the sketch **0.500** inches in each direction, using the sketch plane as the "mid plane."

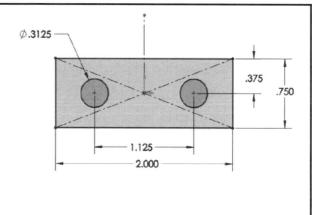

Figure 5-33. Sketch for the base feature of the Spacer.

Now add a new sketch on the top face of the Spacer and **draw** a **Circle** centered at the origin with a <u>diameter</u> of **0.3500** inches, as shown in **Figure 5-34**. Select **Extrude Cut** and cut the hole **Through All** the Spacer.

In the FeatureManager Tree, right click on **Edit Material** and assign **AISI Type A2 Tool Steel** from the SOLIDWORKS Materials – Steel list. Then click the **Apply and Close buttons**. The Spacer is finished. Be sure to **Save** your **SPACER.sldprt** before you **Close** your file.

EYE HOOK

The Eye Hook part can be made by simply revolving a circle and then extruding a circle in their respective planes. Go to your folder and open **ANSI-INCHES.prtdot**. Immediately name your part by pulling down **File** again, select **Save As**, and then name it **EYE HOOK.sldprt**. Select the **Right** plane in the FeatureManager tree, change to a **Right** view orientation. Set the **Units** to **Inches** and **4** decimal places.

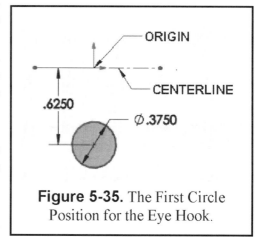

Figure 5-34. Adding the Eye Hook Hole to the Spacer.

Add a new **Sketch** in the **Right** plane and draw a **Circle** in the position indicated in **Figure 5-35**. Also draw a horizontal **Centerline** starting in the origin. Select the **Add Relations** command, and add a **Vertical** relation between the circle's center and the origin. Now use the **Dimension** tool to complete the sketch.

Select the **Revolve Boss/Base** command and revolve the circle **360** degrees. This creates the top part of the Eye Hook.

Select the **Top** plane in the FeatureManager. Pull down **Insert**, select **Reference Geometry**, and then **Plane**.

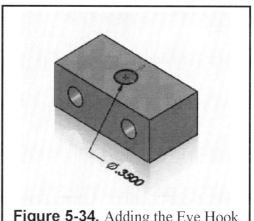

Figure 5-35. The First Circle Position for the Eye Hook.

Define the plane to be **.625** inches **Below** the **Top plane**, and click OK. Now select the new plane (called **Plane1** in the FeatureManager). **Sketch** a **Circle** in the origin with a <u>diameter</u> of **0.3475** inches, as shown in **Figure 5-36**. Next, **Extrude Boss/Base** the circle to a **Blind** distance of **1.2500** inches downward. This completes the Eye Hook as shown in **Figure 5-37**.

In the FeatureManager Tree, right click on **Edit Material** and assign **Tin Bearing Bronze** from the SOLIDWORKS Materials – **Copper Alloys** list. Then click the **Apply and Close buttons**. The Eye Hook now is finished and has a very nice color. Be sure to **Save** your **EYE HOOK.sldprt** before you **Close** your file.

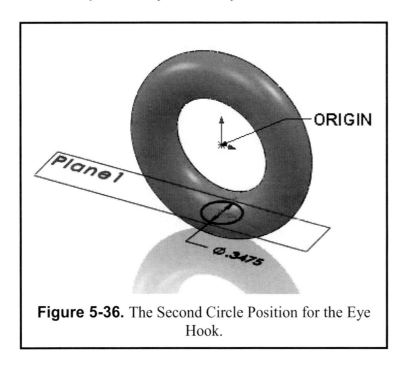

Figure 5-36. The Second Circle Position for the Eye Hook.

Figure 5-37. The Finished Eye Hook.

RIVETS

The final parts to create are the Big and Small Rivets. These are standard parts with dimensions that can be obtained from a handbook. In each case, go to your folder and open **ANSI-INCHES.prtdot.** and **Sketch** the given profile in a **Front** plane. Use the **Dimension** values given in **Figure 5-38** and **5-39** for the Big Rivet and Small Rivet, respectively.

Note: **Do not start the arcs' center for the rivets at the origin.** When you **Dimension** the **Centerpoint Arc** entity, the center of that arc moves down slightly below the origin. Be sure to draw a vertical **Centerline** through the origin. Add a 360 degree **Revolve Boss/Base** to create each Rivet. In the FeatureManager Tree, right click on **Edit Material** and assign **2.0060 (Cu-ETP)** from the SOLIDWORKS DIN Materials – DIN Copper Alloys list. **Save** the Big Rivet as **BIG RIVET.sldprt** before you **Close** your file. **Save** the Small Rivet as **SMALL RIVET.sldprt** before you **Close** your file.

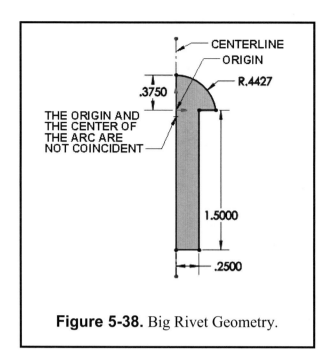

Figure 5-38. Big Rivet Geometry.

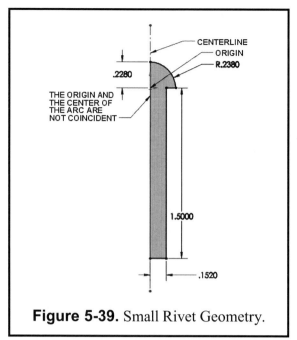

Figure 5-39. Small Rivet Geometry.

SWIVEL EYE BLOCK ASSEMBLY

Select the menu **File, New** and select **Assembly**. From the menu **Tools, Options, Document Properties** and select **ANSI** and then Select **Units**. Make sure you indicate **IPS** (Inch, Pound, Second) and click **OK**. Select the **File** menu again, select **Save As**, and then save it as **SWIVEL EYE BLOCK.sldasm**. Select the menu **File** and **Open,** one-by-one, the four main parts of the assembly:

> **BASE PLATE.sldprt**
> **PULLEY.sldprt**
> **SPACER.sldprt**
> **EYE HOOK.sldprt**

You will add the Rivets later. You now have five files open in SOLIDWORKS (the Assembly file and the four Part files). From the **Window** menu select **Tile Vertically**. You can now see all five files on the screen as shown in **Figure 5-40**.

NOTE: The first part dropped into the assembly is **FIXED** and becomes the base to which all other parts are mated. It cannot be moved unless its constraints are deleted. Starting with the **SPACER** file, select its name in its FeatureManager and "Drag and Drop" it into the assembly's FeatureManager; this way the part's principal planes will be aligned with the assembly's planes. Now "Drag and Drop" the rest of the components **onto the graphics area of the assembly window**, starting with the **PULLEY** and then the **EYE HOOK**. Finally, "Drag and Drop" the **BASE PLATE** <u>two times</u> into the assembly file. You can try to line them up when you "Drop" them into the assembly, but more importantly you will mate them next. Now that the parts have been added to the assembly, you can **Close** their window. Then maximize the assembly window. You should now have an **Isometric** computer screen layout like **Figure 5-41**.

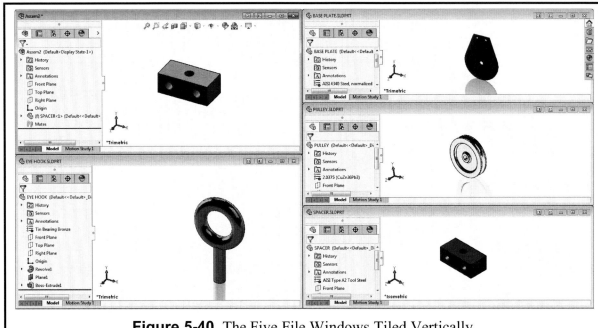

Figure 5-40. The Five File Windows Tiled Vertically.

You will first mate the Front Base Plate with the Spacer. From the **Assembly Tab** select the **Mate** command (it looks like a paper clip), and the "**Mate**" PropertyManager is displayed as indicated in **Figure 5-42**. Now select the upper left small cylindrical hole face of the front Base Plate and the inner left cylindrical hole face of the Spacer. Select the **Concentric** mate and click OK to continue. Repeat this **Concentric** mate process by selecting the upper right small cylindrical hole face of the Base Plate and then select the inner right cylindrical hole face of the Spacer. Use an **Isometric** view and **Zoom** in as needed to facilitate these mates.

Now their holes are concentric but the front Base Plate and Spacer may not touch. Click and drag the Base Plate to confirm.

Figure 5-41. Dragging and Dropping the Parts into the Assembly File.

With the **Mate** command still active select the front face of the Spacer and the back face of the front Base Plate. (*Note:* You probably cannot find a view orientation that allows you to see and select both of these faces simultaneously, so use the **Rotate View** command to select these two faces.) Once you have selected the two flat faces, select the **Coincident** mate and click **OK** button.

You should now see the front Base Plate move towards the front face of the Spacer. The Base Plate is now completely constrained in reference to the Spacer.

Repeat this process with the back side of the Spacer and the second Base Plate. This includes three **Mates**:

- o **Concentric** holes on left side
- o **Concentric** holes on right side
- o **Coincident** faces that touch

The two Base Plates are now mated to the Spacer as shown in an **Isometric** view in **Figure 5-43**.

You can look at it from a **Front** view orientation if you wish to check it and see the holes going all the way through as shown in **Figure 5-44**.

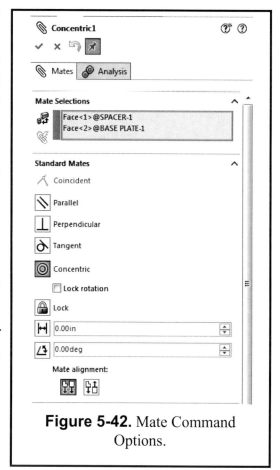

Figure 5-42. Mate Command Options.

Figure 5-43. The Base Plates Mated with the Spacer.

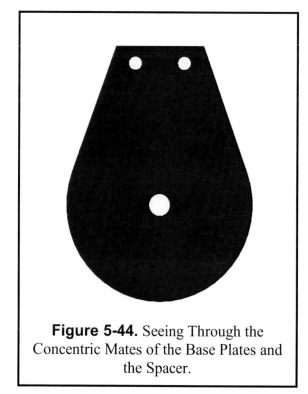

Figure 5-44. Seeing Through the Concentric Mates of the Base Plates and the Spacer.

Now you will mate the Pulley into the Base Plates that will sandwich it. If you finished the **Mate** command, select it again. Now select the large face at the center of the hole in the front Base Plate (*or* the back Base Plate). Next, select the center hole face of the Pulley. Use the **Concentric** mate and then click OK. Notice the Pulley aligns concentrically with the center Base Plate holes but in all likelihood is not sandwiched exactly between them. Switch to a **Right** view orientation to better see this. You could just move the Pulley into position by dragging it, or use the **Move Component** command. However, that is not the right way to do it. You need a second mating condition.

The Spacer is 1.000 inches thick and the pulley is 0.750 inches thick. So you need to position the Pulley (1.000-0.75)/2 = 0.125 inches away from the Base Plates. Select the **Mate** command. Select the front round face of the Pulley and then select the back face of the front Base Plate. You may need to use the **Zoom** and **Rotate View** icons to get these faces selected. The default option is to add a Coincident mate. The faces move until they touch. When this happens select the **Distance** mate, enter a distance value of **0.1250"**, and click OK. The Pulley moves into position and is now at the needed distance. Change to a **Right** view orientation to see the result, as shown in **Figure 5-45**. Make sure there is a small gap. Close the **Mate** command when finished.

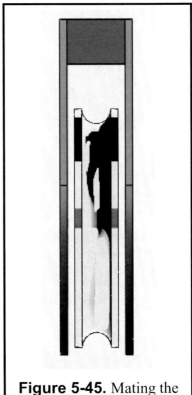

Figure 5-45. Mating the Pulley between the Two Base Plates.

Notice the "Mates" folder in the assembly's FeatureManager. Expanding it will list all the mates made in the Swivel Eye Block Assembly, as shown in **Figure 5-46**. If needed, you can right click on a mate and select **Edit Feature** to modify it.

Now you will mate the Eye Hook to the assembly. Return to an **Isometric** view. Select the **Mate** command. Select the circular shaft of the Eye Hook and the top circular hole on the Spacer. The default option for the selected faces is a **Concentric** mate, and click OK to continue. The shaft now mated with

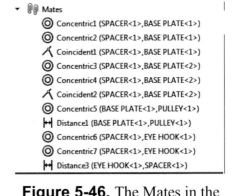

Figure 5-46. The Mates in the Feature Menu.

the hole, but it probably does not go far enough down the hole. Change to a **Right** view and **Zoom** into the area where the shaft goes through the Spacer hole. With the Mate command active, you can click and drag the shaft until it sticks out of the bottom of the Spacer. **Rotate the View** until you can see both the flat bottom of the shaft and the bottom face of the Spacer. Select both the flat bottom of the shaft and the bottom of the Spacer faces. Give them a **Distance** mate of **0.1250** inches. **Preview** the selection (see **Figure 5-47**) and if it is correct, then click OK. If it does not look correct, you can try **Flip Dimension**. Exit the **Mate** command when finished.

The major parts of the assembly are now assembled, and you will add the Rivets next. First though, take note of the degrees-of-freedom that still remain with these parts. The two Base Plates and the Spacer cannot be moved in space because of the two concentric and one coincident mate conditions. However, the Pulley and Eye Hook can still rotate in one degree-of-freedom. Select the **Rotate Component** command on the assembly toolbar, or click and drag the Pulley around its concentric mate and see it turn. Now move the Eye Hook and see how it swivels around in its hole, but you cannot move it out of the hole because of the distance mate. Try to leave the Eye Hook in a position such that the Eye feature is parallel to the Base Plate, as shown in the final **Figure 5-47**.

Figure 5-47. Mating the Eye Hook with the Spacer.

Use the **File, Open** menu and open both the **BIG RIVET** and **SMALL RIVET** parts. From the **Windows** menu select **Tile Vertically**. Drag and drop the "Big Rivet" into the assembly window, and then drag and drop the "Small Rivet" into the assembly window <u>twice</u>. You may want to use the **Move Component**, **Zoom** and **Rotate View** icons to get these faces highlighted simultaneously when you are mating (see **Figure 5-48**). Each rivet will fit into its respective hole and will have the same two identical **Mate** characteristics. The first characteristic will be a **Concentric** mate between the rivet shaft and the inner face of its respective hole. The second mate is a **Coincident** mate between the back flat part of the rivet head and the front face of the front Base Plate. So now execute these mates for the Big Rivet going through the center holes and then for the two Small Rivets going through their respective holes on the top of the assembly. Once you are finished, you should have a final Swivel Eye Block Assembly as shown in **Figure 5-49** in a **Trimetric** view.

Figure 5-48. Dragging and Dropping the Rivets into the Assembly and rotating them.

Manufacturing Note: If you **Rotate View** and look at the back of your assembly, you can see the Rivets sticking out about 0. 25 inches (one fourth of an inch). The normal process used to secure the Rivet to the Base Plate is a peen operation. *Peening* the end of the Rivet with a hammer, for example, causes the face to expand in a spherical direction, and thus prevents the Rivet from coming out of the hole. We will leave this peening operation for a later day, and your assembly modeling and mating exercise is over.

Pull down **File**, select **Save As**, and save the assembly as **SWIVEL EYE BLOCK.sldasm** in your folder.

Before you close this file, you should save it and also get a hard copy printout. Insert a shaded Isometric view of the Swivel Eye Block Assembly into your **Title Block** drawing sheet.

You will now create a Bill of Materials and add Balloons with Part Names attached to the assembly. To begin, go to Tools – Options – Document Properties and make the changes that are shown in **Figure 5-27**. Set Leader style and Frame style to **0.0098in**. The leader display for Single/Stacked Balloons and Auto Balloons should both be bent. Under the Single balloon selection box, set the style to Circular and the size to 2 Characters. Now click on **Font**. And set the Font to **Arial – Regular** – and the **Height to 0.25in**. Once all of these settings are made, click the **OK** tab.

Figure 5-49. The Finished Swivel Eye Block Assembly in an Isometric View.

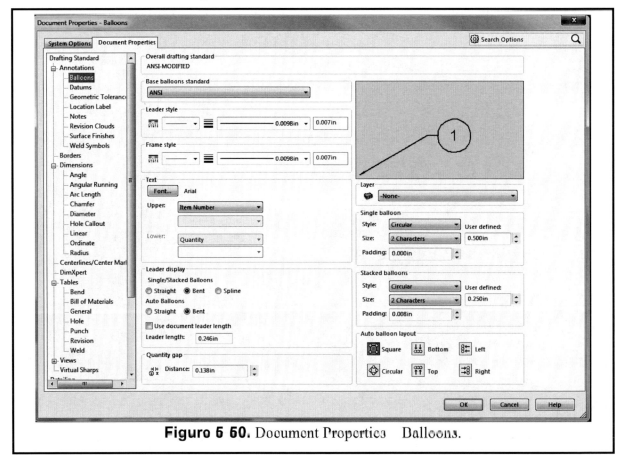

Figure 5-50. Document Properties – Balloons.

Select **Insert–Annotations – Balloon** and carefully select the Spacer, one Base Plate, the Pulley and Eye Hook, one Small Rivet and the Big Rivet. Circles with the respective Part numbers will appear. After all of the Balloons are placed, select **Insert Annotations Note** (without a Leader) and place the part name (18 pt) next to the corresponding balloon.

Now select **Insert – Tables – Bill of Materials.** Once you select the assembly, a selection menu will appear where the FeatureManager tree is. Select Parts only, Display all configurations of the same part as one item; for the border, the first number should be changed to 0.0098 and the second number is unchanged. Okay your selections and a Bill of Materials will appear on the drawing sheet. Place it in the upper right corner of your drawing.

Now save your drawing as **PULLEY ASSEMBLY.slddrw** and **Print** it on this sheet (see **Figure 5-51**). If you have access to a *Color Printer*, use it to show the nice colors of your assembly.

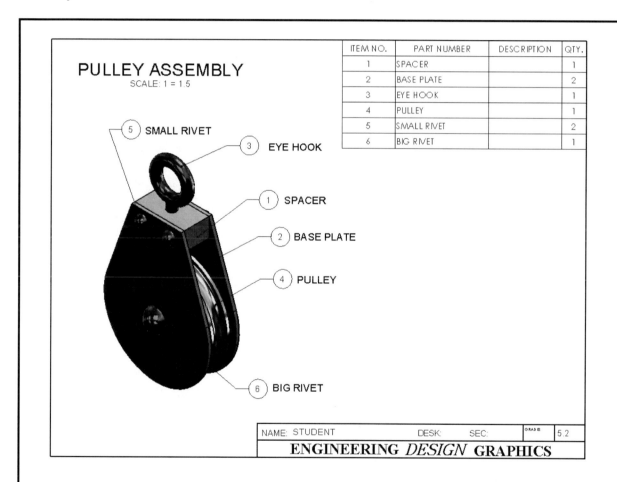

ITEM NO.	PART NUMBER	DESCRIPTION	QTY.
1	SPACER		1
2	BASE PLATE		2
3	EYE HOOK		1
4	PULLEY		1
5	SMALL RIVET		2
6	BIG RIVET		1

PULLEY ASSEMBLY
SCALE: 1 = 1.5

5 SMALL RIVET
3 EYE HOOK
1 SPACER
2 BASE PLATE
4 PULLEY
6 BIG RIVET

NAME: STUDENT   DESK: SEC: GRADE   5.2
ENGINEERING *DESIGN* **GRAPHICS**

Figure 5-51. The Swivel Eye Block Assembly Isometric view in the Title Sheet.

Be sure to save your assembly files for use in later lab exercises such as "Kinematics Simulation" where you will need to explode an assembly and make an animation file.

SUPPLEMENTARY EXERCISE 5-1: CASTER ASSEMBLY

Create the parts of the Caster Assembly of the Option specified by your instructor in the Table provided.

1. Generate an Assembly Drawing of the Caster Parts and insert the Assembly into a Title Block.

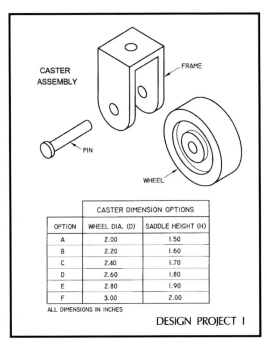

CASTER ASSEMBLY

FRAME

PIN

WHEEL

	CASTER DIMENSION OPTIONS	
OPTION	WHEEL DIA. (D)	SADDLE HEIGHT (H)
A	2.00	1.50
B	2.20	1.60
C	2.40	1.70
D	2.60	1.80
E	2.80	1.90
F	3.00	2.00

ALL DIMENSIONS IN INCHES

DESIGN PROJECT 1

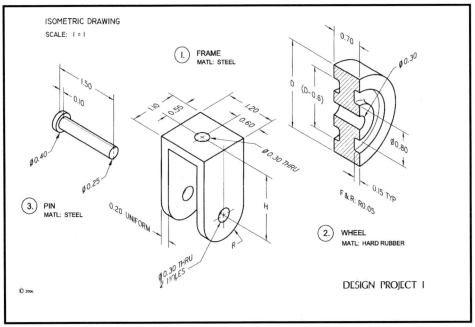

ISOMETRIC DRAWING
SCALE: 1 = 1

1. FRAME
MATL: STEEL

3. PIN
MATL: STEEL

2. WHEEL
MATL: HARD RUBBER

DESIGN PROJECT 1

SUPPLEMENTARY EXERCISE 5-2: PULLEY ASSEMBLY

Create the parts of the Pulley Assembly of the Option specified by your instructor in the Table provided.

1. Generate an Assembly Drawing of the Pulley Parts and insert the Assembly into a Title Block.

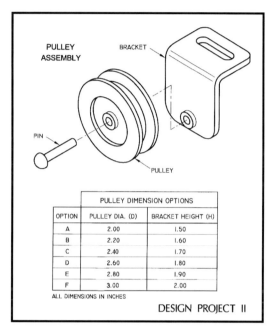

OPTION	PULLEY DIA. (D)	BRACKET HEIGHT (H)
A	2.00	1.50
B	2.20	1.60
C	2.40	1.70
D	2.60	1.80
E	2.80	1.90
F	3.00	2.00

ALL DIMENSIONS IN INCHES

DESIGN PROJECT II

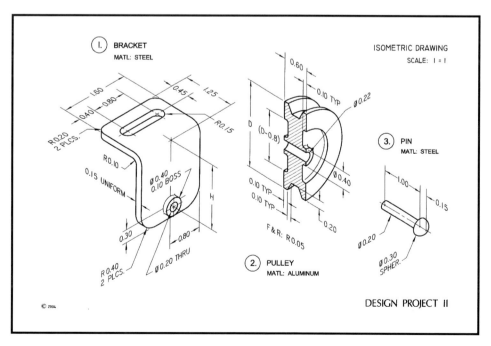

NOTES:

Computer Graphics Lab 6: Analysis and Design Modification I

In Computer Graphics Lab 6, you will be introduced to some of the advanced evaluation and design modification capabilities of SOLIDWORKS. Solid modeling is a tool for engineering design that offers many advantages over conventional practices. One of these advantages is the ability to analyze the design directly from the digital database without the need to build a physical part. In this laboratory, you will evaluate various design properties of the solid model using the **Measure** and **Mass Properties** commands. You will also be exposed to **Design Tables,** which can be used to build a family of similar parts.

THE MEASURE TOOL

The **Measure** tool can be found in the **Tools** pull-down menu (see **Figure 6-1**). When you use this function, a ruler appears on the cursor and you can select the entity (entities) to measure in the screen. This function includes the following capabilities as examples.

They can be used in a 2D sketch or a 3D solid model.

Line returns the length of the line.

Arc returns the length of the arc and the diameter. If the arc is a full circle, the length ends up being the circumference (see **Figure 6-2**).

Model Face returns the area and the total length of all model edges connected to the face.

Figure 6-1. The Measure Function.

THE MASS PROPERTIES FUNCTION

From the **Tools, Evaluate** menu, select the **Mass Properties** command (see also **Figure 6-1**). This will open an on-screen window that reports the mass properties of the current solid model. Before you can calculate the mass properties accurately, you need to set the density of the material for the solid model part. The density is set when the material is assigned. In the FeatureManager Tree right click on "Material" and select Edit Material. Once the material is assigned the density of the material is also set, as shown in **Figure 6-3**. See also **Table 6-1** for some common density values.

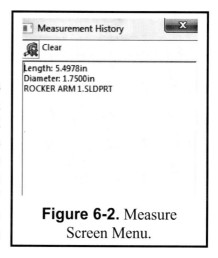

Figure 6-2. Measure Screen Menu.

Once the density of the part material is set, you can then select the **Mass Properties** command. The result will be reported in the "Mass Properties" Report in the on-screen window, as shown in **Figure 6-4**. The mass properties and their units of measure reported include the following (see **Page 6-11**, for definitions):

- **Part Name**
- **Density (lbs/in³⁾**
- **Mass (lbs)**
- **Volume (in³)**
- **Surface Area (in²)**
- **Center of Mass (in)**
- **Moments of Inertia (lbs*in²)**

After the component's "Mass Properties" are calculated, you can review the results in this window. Keep in mind that this window can be resized if needed. The options available include:

Print will open the print dialog where you can print a copy of the "Mass Properties" Report.

Copy will copy the entire report to the computer clipboard. Then you can paste this information directly into other documents. This allows you to edit the report or customize it.

Options will allow you to change the units of measure of the mass properties calculated, shown in **Figure 6-5**.

Recalculate will recalculate the mass properties after other options are changed.

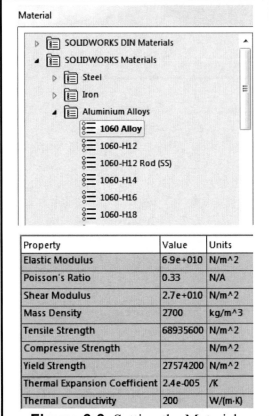

Figure 6-3. Setting the Materials Properties.

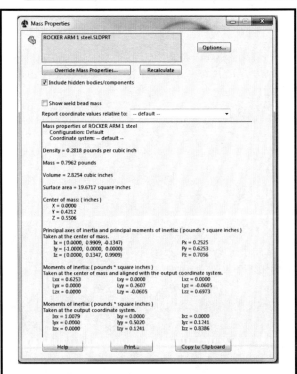

Figure 6-4. Mass Properties Report.

ABOUT THE MASS PROPERTIES UNITS

Some mass properties are based on the geometric distribution of the model and are independent of the material's density (assumed to be uniform). Examples of these properties are volume and center of mass (centroid). Other properties, such as mass and moments of inertia, are dependent on the material's density. **Table 6-1** lists the densities of some common engineering materials. You can use these density values to set the material properties of your part. Weight and Mass are often confused. Below is an example of the correct way to calculate them for a unit one-inch cube.

Note: The "Density" value used by SOLIDWORKS (see **Figure 6-5**) is actually the value of the "Unit Weight" listed below in **Table 6-1**. This lack of uniformity in terminology contributes to the general confusion on this matter.

Figure 6-5. Mass Property Options.

Example Calculation: Material is Mild Steel (density is 0.728×10^{-3} lbs-sec^2/in^4)

Weight = (unit weight)x(volume) = (0.281 lbs/in^3) x (1.00 in^3) = 0.281 lbs.

Mass = (density)x(volume) = (0.728×10^{-3} lbs-sec^2/in^4)x(1.00 in^3) = 0.728×10^{-3} lbs-sec^2/in.

= (weight) / (gravity) = (0.281 lbs) / (386 in/sec^2) = 0.728×10^{-3} lbs-sec^2/in.

Table 6-1. Some Unit Weights and Densities of Common Materials.

Material	Unit Weight (lbs/in^3)	Density (lbs-sec^2/in^4)
Aluminum	0.097	0.251×10^{-3}
Brass	0.307	0.794×10^{-3}
Chromium	0.259	0.671×10^{-3}
Copper	0.323	0.837×10^{-3}
Magnesium	0.063	0.163×10^{-3}
Plastic	0.036	0.093×10^{-3}
Rubber	0.041	0.106×10^{-3}
Steel	0.281	0.728×10^{-3}
Titanium	0.163	0.422×10^{-3}

Exercise 6.1: ROCKER ARM MASS PROPERTIES

For Exercise 6.1, you will build the basic geometry of the rocker arm, which is designed to rotate about a principal axis. You will then copy the first rocker arm data into a second rocker arm file and make a change in its geometry. Mass Properties calculations will be performed on both models and the results will be compared to each other.

ROCKER ARM 1

Open **ANSI-INCHES.prtdot** and **Save** it as **ROCKER ARM 1.sldprt**. In the **Front** plane, **Sketch** a **Circle** with a <u>diameter</u> of **1.75** inches and centered at the **ORIGIN**. **Extrude** it a **Blind** distance of **1.25** inches **OUT** from the front plane. You now have the bottom Boss/Base.

Add a new **Sketch** in the **Front** plane and draw the back upright profile of the Rocker Arm as indicated in **Figure 6-6**. Draw the three **Lines** such that the bottom of the profile slightly extends into the extruded base. Draw the outer top arc using the **Tangent Arc** tool. Draw the **Circle** concentric to this arc. Now **Add Relations** to the profile as follows:

- o The center of the small circle and the origin are **Vertical**.
- o The small circle and top arc are **Concentric**.
- o The left-side line and top arc are **Tangent**.
- o The right-side line and top arc are **Tangent**.

Finally, use the **Dimension** tool to add the three dimensions given in **Figure 6-6**. Now **Extrude** this upright sketch **0.75** inches **OUT** from the front plane. You will get the beginning model of the Rocker Arm as shown in **Figure 6-7**.

Now you will draw the final sketch to cut the hole and keyway through the bottom boss. Select the front face and add a new **Sketch**. Change to a **Front** view orientation and draw the big **Circle** with a <u>radius</u> of **0.50** inches centered at the origin. Draw three short **Lines** to make the keyway at the top of the circle and then use the **Trim** tool to cut away the segments that are not needed. Refer to **Figure 6-8** to see how it should look at this point. Use the **Dimension** tool to dimension the sketch as shown in **Figure 6-8**.

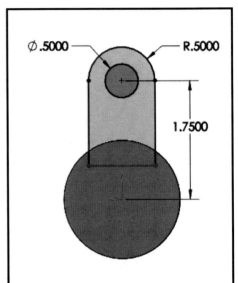

Figure 6-6. Dimensions for the Upright Part of the Rocker Arm.

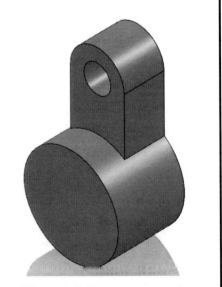

Figure 6-7. The Beginning Rocker Arm.

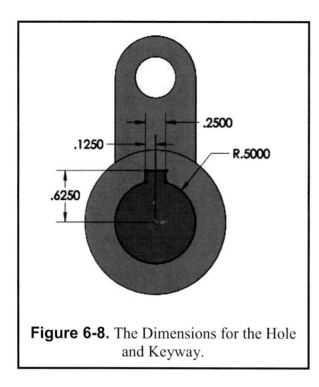

Figure 6-8. The Dimensions for the Hole and Keyway.

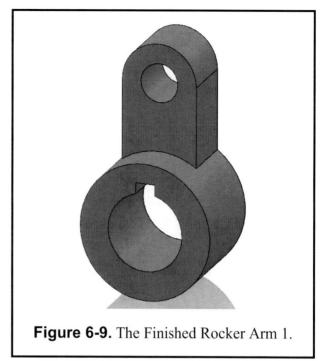

Figure 6-9. The Finished Rocker Arm 1.

With the sketch finished, make an **Extruded Cut** using a **Through All** end condition. The Rocker Arm 1 is finished as shown in **Figure 6-9** in a **Dimetric** view. **Save** your model as **ROCKER ARM 1**. Do <u>not</u> close your file.

ROCKER ARM 2

The Rocker Arm 2 will be very similar to Rocker Arm 1. Use the **Save As** command and save the model as **ROCKER ARM 2.sldprt**. The three differences for the new Rocker Arm 2 are the width of the upright feature (now 1.25 inches) the height from the origin to the center of the circle (1.50 inches) and the depth of this upright feature (now only 0.50 inches). These changes can be accomplished easily using the feature editing capabilities of SOLIDWORKS.

In the FeatureManager tree, expand the **Boss-Extrude2** feature, select **Sketch2** and click on **Edit Sketch** from the context toolbar. Now we are editing the second sketch. Double click to change the dimensions as needed of the large arc from R0.500 to **R0.625** and the height dimension from 1.75 to **1.500** as shown in **Figure 6-10**. Then click on the **Rebuild** command to update the model with the new dimensions.

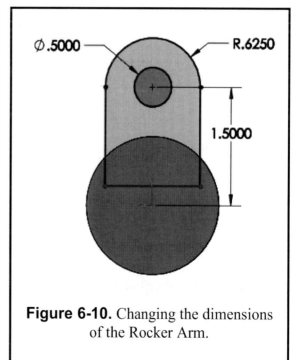

Figure 6-10. Changing the dimensions of the Rocker Arm.

Now change the extrusion depth of the upright feature. Select the second "Boss-Extrude" feature in the FeatureManager tree. Select the **Edit Feature** option and change the extrusion's depth to **0.500** inches. Then click **OK** to rebuild the model. The solid model will now be rebuilt with the new extrusion depth. The Rocker Arm 2 is now finished as shown in **Figure 6-11** in a **Trimetric** view. **Save** your model as **ROCKER ARM 2**. Then **Close** your SOLIDWORKS file to start the next phase of this analysis exercise.

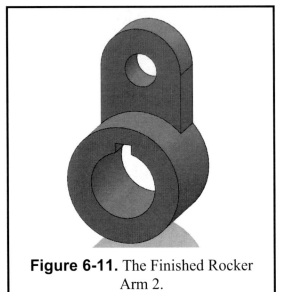

Figure 6-11. The Finished Rocker Arm 2.

MASS PROPERTIES ANALYSIS OF ROCKER ARM 1

You will start your mass properties analysis with Rocker Arm 1. **Open** the **ROCKER ARM 1** file from your folder. Under **Tools – Options – Document Properties – Units - Mass/Section Properties** change **Length** to **4 decimal places**. Now you need to set the material for the Rocker Arm 1.

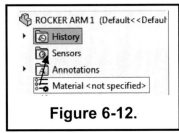

Figure 6-12.

Right click on the **Material** feature (**Figure 6-12**) and select **Edit Material**. Under **SOLIDWORKS Materials** in the "Materials Editor" menu (**Figure 6-13**), expand the **Steel** library of materials, select **CAST CARBON STEEL** and click on **Apply** and then **Close**.

From the menu **Tools, Evaluate** select the **Mass Properties** command. The "Mass Properties" report window will now appear on the screen showing all the mass properties for Rocker Arm 1 made of cast carbon steel. The "Mass Properties" window will be displayed as shown in **Figure 6-14**.

Now we want to copy the contents and personalize the "Mass Properties" report. So, in the "Mass Properties" window, click the **Copy to Clipboard** button. This will copy the entire report to the computer's clipboard. Now open a word processing software like MS Word. Use the **Edit** and **Paste** commands in the software to paste the clipboard data directly onto a new page. At the top of the Mass Properties report, type in Your **Name,** Your **Seat** Number, and Your **Section** Number.

Figure 6-13. Materials Editor Menu.

Also add an opening sentence of the Mass Properties report: "**Mass Properties of Rocker Arm 1 made of Cast Carbon Steel**" in order to identify the part name and assigned material. Next, select **File** and **Save** in the word processing software to save the word file as **ROCKER ARM 1 - STEEL**. Then close your word processing software and return to SOLIDWORKS. You will print this document after you determine which of the mass property reports satisfies the question at the bottom of **Page 6-8**.

When you return to SOLIDWORKS, the "Mass Properties" window is still open, so click the **Close** button. You will now repeat this mass properties analysis for Rocker Arm 1 using **Aluminum** as the material.

Right click on the **Material** (currently Cast Carbon Steel) icon (**Figure 6-12**) in the FeatureManager. Under **SOLIDWORKS Materials**, expand the **Aluminum** library of materials and select **1060 ALLOY**. This sets all the mass properties for this part, then click on **Apply** and then **Close.**

From the menu **Tools, Evaluate** select the **Mass Properties** command. The "Mass Properties" report window will now appear on the screen showing all the mass properties for Rocker Arm 1 made of aluminum material, as shown in **Figure 6-15**. Now repeat the **Copy** to clipboard and **Paste** in a word processing software procedure used previously. Be sure to include your pertinent data and material selection. **Save** your **ROCKER ARM 1 - ALUMINUM** file and then you can close the word processing software. You will print this document after you determine which of the mass property reports satisfies the question at the bottom of **Page 6-8**.

When you return to SOLIDWORKS, **Close** your "Mass Properties" window, and **Close** the **ROCKER ARM 1** part file to start analysis on Rocker Arm 2.

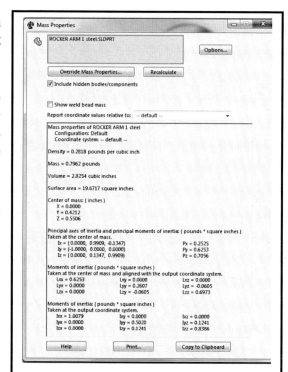

Figure 6-14. Mass Properties for Cast Carbon Steel Rocker Arm 1.

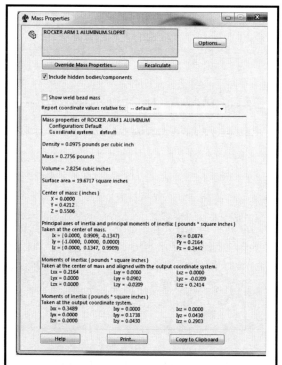

Figure 6-15. Mass Properties for Aluminum Rocker Arm 1.

MASS PROPERTIES ANALYSIS OF ROCKER ARM 2

Now **Open** the **ROCKER ARM 2** file and repeat this mass properties analysis with the Rocker Arm 2 solid model. Repeat all the previous steps, first for **Cast Carbon Steel** and then for **1060 Alloy Aluminum**. Get word files of these two reports using the copy and paste procedure outlined previously. Be sure to type your name, class data and material on the reports before you **SAVE** them. Refer to **Figures 6-16** and **6-17** as needed for these mass property reports. You can **Save** your reports as files named **ROCKER ARM 2 - STEEL** and **ROCKER ARM 2 - ALUMINUM**. When you are finished, **Close** down your **ROCKER ARM 2** part file. You should now have reports of various mass property reports saved. You will print these documents after you determine which of the mass property reports satisfies the question at the bottom of **Page 6-8**.

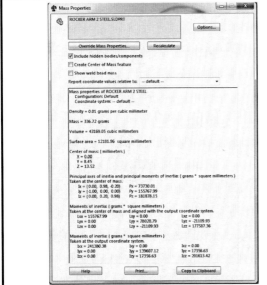

Figure 6-16. Mass Properties for Cast Carbon Steel Rocker Arm 2.

Figure 6-17. Mass Properties for Aluminum Rocker Arm 2.

COMPARISON OF MASS PROPERTIES FOR ROCKER ARMS 1 and 2

After you have completed the mass properties analysis for both Rocker Arms 1 and 2, you can make some comparisons. Because of the design and functionality of the rocker arm, a key parameter is motion about the central axis of the large cylindrical hole on the bottom. In this case, it would be motion about the Z-axis. Hence, the moment of inertia (resistance to steady rotation) about the Z-axis taken at this center would be considered a critical physical property. Study the moment of inertia about the Z-axis of the two rocker arm designs for each of the two assigned materials (mild steel and aluminum), and answer the question below on the correct Mass Property report.

> **Note:** You now have four Mass Properties reports. Identify on the correct report, which combination of geometry and material yields a Rocker Arm design that is easiest to rotate (requires the least torque) about the central Z-axis?

CREATING A SHADED IMAGE OF THE TWO ROCKER ARMS

You will now create shaded images of the two Rocker Arms, side-by-side, and obtain a plot for submission to your instructor. You will do this using an assembly model. Pull down **File**, select **Open**, and then select the **ROCKER ARM 2** part file.

Repeat this **File**, **Open** sequence for **ROCKER ARM 1** also. Now pull down **File** and select **New** - **Assembly**. This will create a blank assembly file. Pull down **File**, select **Save As**, and then give your new assembly file the name **ROCKER ARM ASSEMBLY**. At first, all of these computer screen images (viewports) land on top of each other. To see all three screens, pull down **Window** and select **Tile Vertically**. This will result in a screen image as shown in **Figure 6-18**.

To assemble your two Rocker Arms, just pick, drag, and drop your parts into the assembly window. Click on the **ROCKER ARM 1** icon in its FeatureManager window. Drag it into the assembly window and drop it into an area around the origin. In like manner, go to the **ROCKER ARM 2** icon in its FeatureManager window and drag and drop it into an area just to the right of Rocker Arm 1. Both parts are now in the assembly. You no longer need the two Rocker Arm files, so you can close both of those windows. Also, you can now maximize your Rocker Assembly window.

The Rocker Arms may have been dropped into somewhat arbitrary positions. So first try a **Trimetric** View Orientation. Now arrange Rocker Arm 2 to be on the right side of Rocker Arm 1. Select the **Move Component** icon (refer back to **Figure 5-3**) from the Assembly Toolbar. With the move cursor, drag Rocker Arm 2 to move it to an acceptable position (see **Figure 6-19** as a guide). You can also use **Zoom** and **Pan** to arrange your image. You may now save your assembly, so pull down **File** and select **Save As**. Save your **ROCKER ARM ASSEMBLY**.

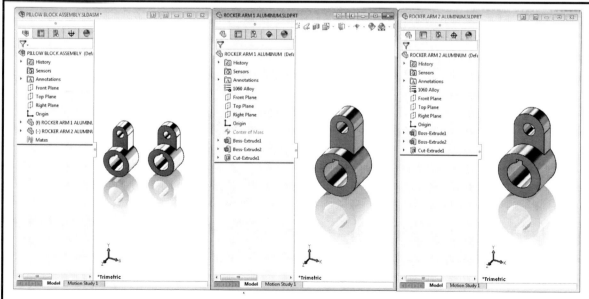

Figure 6-18. The Rocker Arm Assembly, Rocker Arm 1, and Rocker Arm 2 Windows Tiled Vertically.

Open the **TitleBlock-Inches.drwdot** from your folder and insert a dimetric view of the Assembly in the same manner that you have been doing with single solid parts. Before you print a hardcopy of your image, you need to label the Rocker Arms. Pull down the **Insert** menu, select **Annotations**, and then select **Note**. The on-screen "Note" menu appears. Type in the label "**ROCKER ARM 1**" in the note window. You may want to dictate the font size also, so click **off** the check mark in the "Use Documents Font window." Click the **Font** button and the "Choose Font" on-screen menu appears. Here you can set the font type and height. Use the **Arial** font type with a **Bold** setting and set the height to **0.20** inches. Then click **OK** twice to close the on-screen menus, and the note appears on the screen. In all likelihood, the note is not positioned correctly, so drag it with the LMB to an appropriate position above **ROCKER ARM 1**. Repeat this annotation process for **ROCKER ARM 2**. Refer to **Figure 6-19** for an example of an acceptable screen layout.

Now preview your print file by pulling down **File** and select **Print Preview**. See your image as it will appear when printed as shown in **Figure 6-19**, and then click the **Close** button. Then pull down **File** and select **Print**. Make sure the proper printer is selected, and then click **OK**. This finishes the Lab Exercise 6.1 and you can pull down **File** and click **Close** to exit.

Note: Attach your Rocker Arm Assembly printout to your four Mass Properties reports and submit them to your instructor before you leave the lab.

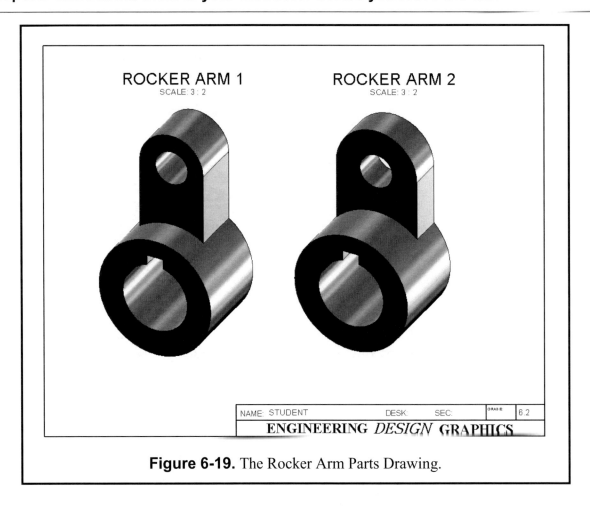

Figure 6-19. The Rocker Arm Parts Drawing.

INFORMATION PROVIDED

IN

SOLIDWORKS MASS PROPERTIES REPORTS

1. **DENSITY** is the mass or the weight per unit volume of the material the part is made from.

2. **MASS**: The mass of a body is the measure of its property to resist change in its steady motion. The mass depends on the volume of the body and the density of the material of which the body is made. In this case with SOLIDWORKS, mass is equivalent to weight.

3. **VOLUME**: The volume of a body is the total volume of space enclosed by its boundary surfaces.

4. **SURFACE AREA**: The surface area is the total area of the boundary surfaces defining the solid model.

5. **CENTER OF MASS**: Center of Mass (or Centroid) of a volume is the origin of coordinate axes for which first moments of the volume are zero. It is considered the center of a volume. For a homogeneous body in a parallel gravity field, mass center and center of gravity coincide with the centroid.

6. **PRINCIPAL AXES OF INERTIA AND PRINCIPAL MOMENTS OF INERTIA**: Principal moments of inertia are extreme (maximal, minimal) moments of inertia for a body. They are associated with principal axes of inertia which have their origin at the centroid and the direction of each usually given by the three unit-vector components. For these axes, the products of inertia are zero.

7. **MOMENTS OF INERTIA**: A moment of inertia is the second moment of mass of a body relative to an axis, usually X, Y, or Z. It is a measure of the body's property to resist change in its steady rotation about that axis. It depends on the body's mass and its distribution around the axis of interest.

Exercise 6.2: SOCKET DESIGN TABLE

In Exercise 6.2, you will make a family of parts of a Socket using a Design Table. A Design Table allows you to create multiple configurations (or variations) of similar parts or assemblies, by specifying parameters in an embedded Microsoft Excel worksheet. These parameters include the dimensions of the model that will vary from one configuration to the next. Once a design table is added, you can go to the "ConfigurationManager" to review the different configurations, or variations, and can be used in assemblies and drawings.

Open **ANSI-INCHES.prtdot** and **Save** it as **SOCKET.sldprt**. Start a new **Sketch** in the **Top** plane. Draw a **Circle** and dimension it as shown in **Figure 6-20**.

When you add dimensions, SOLIDWORKS automatically assigns them an internal name with the format:

DimensionName @ FeatureName

The **DimensionName** is the name of a dimension; the automatically assigned names have the format D1, D2, etc. for each sketch or feature. **FeatureName** is the name of the Feature or Sketch the dimension belongs to. For example, the **0.770"** dimension was the first dimension added to *Sketch1*; therefore, it will be named "*D1@Sketch1*."

A Design Table uses the SOLIDWORKS' dimension names to assign different dimension values for each configuration.

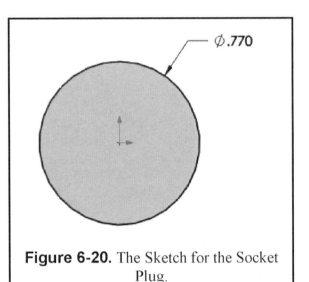

Figure 6-20. The Sketch for the Socket Plug.

When you select a dimension, its name and other properties are listed in the dimension's PropertyManager. Select the **0.770 in** dimension. Its properties are shown in **Figure 6-21**. In the "Primary Value" section the dimension's name is listed as **D1@Sketch1** and shows the value of that dimension (**0.770 in**).

To make it easier to identify dimensions used in a Design Table, we will rename them. With the **0.770 in** dimension selected, type "**Socket_Diameter@Sketch1**" in the name Text Box, replacing the old name, *D1@Sketch1*. The resulting dimension name is shown in **Figure 6-22**.

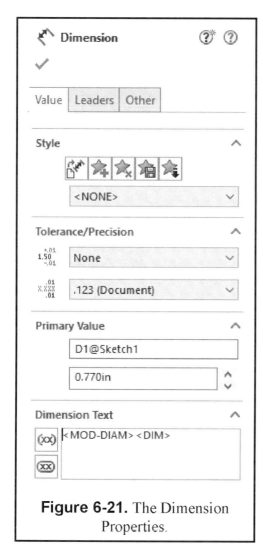

Figure 6-21. The Dimension Properties.

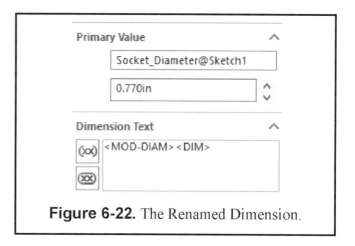

Figure 6-22. The Renamed Dimension.

When you are finished renaming the dimension, select the **Base-Extrude** command and extrude the sketch **0.960 in** upward to create the base feature of the Socket, as shown in **Figure 6-23**.

When the Boss-Extrude1 is created, its depth is also

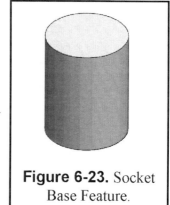

Figure 6-23. Socket Base Feature.

assigned a dimension. To make the model dimensions visible you can right-click in the **Annotations** Folder of the FeatureManager and activate the **Show Feature Dimensions** option, making all of the model dimensions visible, until the option is turned off.

Select the **0.960 inch** dimension, and type the name:
"**Socket_Height@Boss-Extrude1**" in the dimension's properties. "Boss-Extrude1" is the name of the extrusion. The result is shown in **Figure 6-24**. Right click in the **Annotations** folder and turn off the **Show Feature Dimensions** option.

From the **Features Tab** select the **Fillet** command. Add a **.03 inch** fillet to the top edge of the **Base Part** as shown in **Figure 6-25.**

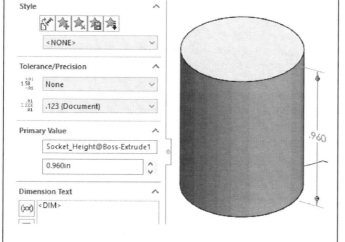

Figure 6-24. The Boss-Extrude1 Dimension Properties.

Now you will create the square hole in the bottom of the socket where the ratchet wrench fits. Change to a **Bottom** view orientation, and select the **Bottom** face of the part. Add a new **Sketch** and draw a **Rectangle** as shown in **Figure 6-26**.

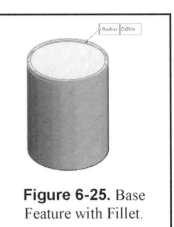

Figure 6-25. Base Feature with Fillet.

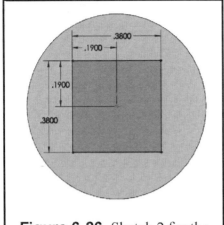

Figure 6-26. Sketch 2 for the Ratchet Wrench Opening.

Use the **Dimension** tool to add the dimension values indicated. Next, make an **Extruded Cut** feature to a depth of **.4500 inches**.

Change to a **Top** view, select the top face of the Socket and add a new Sketch. Draw a **hexagon** with an **inscribed circle** diameter of **.5660 in.** at **0 degrees**. Even though it is easy to define the hexagon's diameter of **.5660 in** in the PropertyManager, it is not a dimension that you can use in a design table. You need a dimension in the sketch for use in the Table. Select the **Smart Dimension** tool and dimension the inscribed circle to **.5660 inches**. Now select the diameter dimension and change its name to **Socket_Size@Sketch3**, as shown in **Figure 6-27**. Make an **Extruded Cut** using the **Up To Surface** end condition and **select** the **top** face of the **square hole** on the bottom of the Socket, as shown in **Figure 6-28**.

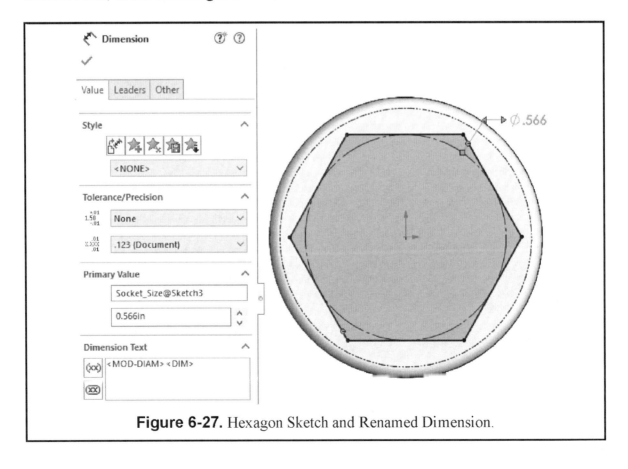

Figure 6-27. Hexagon Sketch and Renamed Dimension.

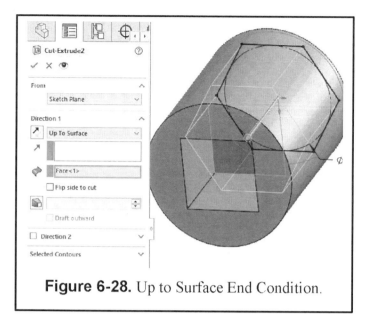

Figure 6-28. Up to Surface End Condition.

To complete the socket opening, from the **Features** tab select the **Circular Pattern** command, select a circular edge of the first extrusion as the direction, and select the hexagonal cut in the **Features to Pattern** selection box. Use the "Instance Spacing" option, enter **30 degrees** and **2** copies, as shown in **Figure 6-29**, and click **OK** to finish. The completed socket opening is shown in **Figure 6-30**.

Figure 6-29. Circular Pattern parameters for the Second Hexagon.

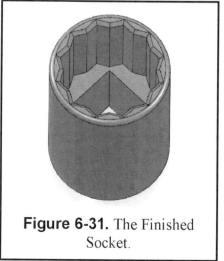

Figure 6-31. The Finished Socket.

Figure 6-30. The Finished Socket Opening.

The last step in the construction of the Socket is to add a **Chamfer** to all of the short edges of the polygon on the top face of the Socket. Set the value of the **Chamfer** to **.023 in**. The finished socket should look like the socket in **Figure 6-31**. Save your model as **SOCKET.sldprt**.

You are ready to add a Design Table. From the **Insert** menu, select **Tables – Design Table**. Select **Auto-create** on the Design Table properties, and click **OK** to continue. You will see a **Dimensions** window, listing all the dimensions available in the model. Hold down the **Ctrl** key, select the following dimensions to add them to the Design Table, and click **OK** to continue:

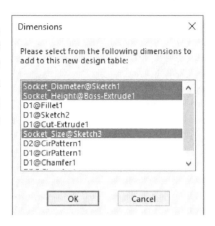

- **Socket_Diameter@Sketch1**
- **Socket_Height@Boss-Extrude1**
- **Socket_Size@Sketch3**

Now an Excel spreadsheet is added to the SOLIDWORKS part, as seen in **Figure 6-32**. The selected dimensions have been added to the "Design Table" starting on Row 2, Column B. To add configurations to the Design Table, you have to enter the configuration names starting on Row3, Column A, and additional configurations going down, then fill the table with the corresponding dimension values for each configuration. Fill the Design Table as shown in **Figure 6-32.**

Note: **Be careful to not click outside of the spreadsheet boundary, because when you click outside of it, the Design Table is considered complete and you will exit Excel and return to SOLIDWORKS.** When you finish entering the table information, click on the **SOLIDWORKS** graphics area to complete the Design Table and create the different configurations.

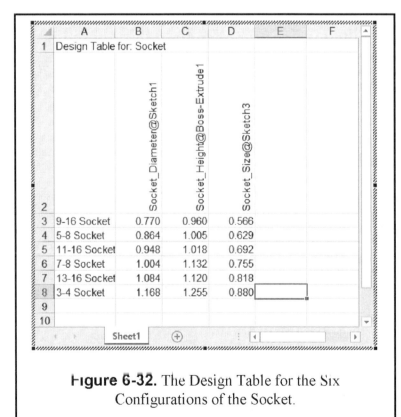

Figure 6-32. The Design Table for the Six Configurations of the Socket.

If you need to modify your design table after completing it, select the Configuration Manager tab at the top of the FeatureManager **(Figure 6- 33)**, expand the Tables folder, right click on **Design Table** and select **Edit Table**.

Figure 6-33. The Configuration Manager Tab.

After you click in the graphics area, the configurations are generated and are listed as shown in **Figure 6-34**. If needed, you can Right click and select **Edit Table** of the Design Table to make changes, add or delete configurations.

Note: Sometimes when you are completing a Design Table, you may make a mistake or inadvertently close it before you are finished. Right click on the **Design Table** in the **ConfigurationManager** and select **Edit Table** to modify it (see **Figure 6-34**).

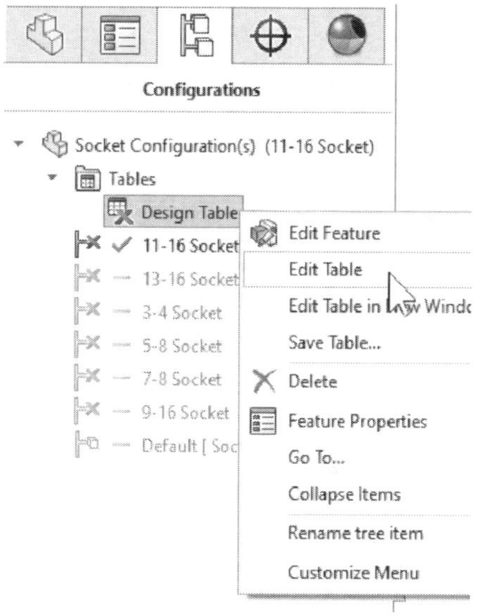

You are now ready to see the six different design configurations created by the Design Table. You might also want to **Save** your part file now to be safe.

Select the FeatureManager tab, indicated in **Figure 6-33**. The configuration name with a green check mark is the currently active configuration. In an **Isometric** view, one-by-one, double click on the grayed configuration names to display them. Double click in each configuration, **11-16 Socket, 13-16 Socket, 3-4 Socket, etc.** to see their differences. You may be asked to **Rebuild** your model during this process. Notice the configurations driven by the **Excel** table are preceded by an **Excel** icon, and the **Default** configuration is not. Also, the **Default** configuration happens to be the same as "**9-16 Socket**" because the values entered in row 3 are exactly the same as your original model dimensions. *Note:* Since the "**9-16 Socket**" and the "**Default**" configuration have the same dimensions, the **Default** configuration can be deleted. **Save** your file again.

Since this is a family of parts, it would be nice to view them all together to see their variations in size and shape. One way to do this is to "drag and drop" them all into an assembly. From the **File** menu select **New,** and select **Assembly**. Pull down **Window** and **Tile Vertically**. Now "drag and drop" the configuration names from the ConfigurationManager one-by-one into the assembly window in the following order:

- **9-16 Socket**
- **5-8 Socket**
- **11-16 Socket**
- **7-8 Socket**
- **13-16 Socket**
- **3-4 Socket**

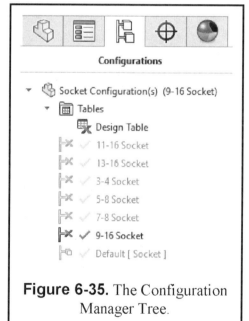

Figure 6-34. Edit Design Table.

Figure 6-35. The Configuration Manager Tree.

You can now maximize the assembly window to see the operation better. Arrange them in the assembly window using the **Move Component** command to get a layout as suggested in **Figure 6-36** without the labels. Once aligned, save the assembly. Open **TITLEBLOCK-INCHES** and insert your assembly view the same way you have been inserting your part views. From the **Insert, Annotation, Note** menu add the titles and **Print** the drawing to hand to your instructor. **Save** your file as **SOCKET ASSEMBLY.slddrw** and close it.

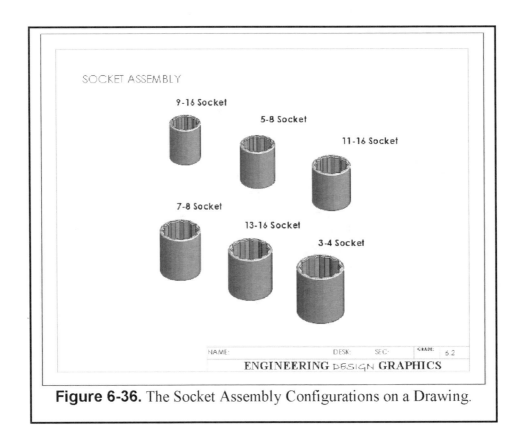

Figure 6-36. The Socket Assembly Configurations on a Drawing.

SUPPLEMENTARY EXERCISE 6-1: CASTER AND CASTER FRAME DESIGN TABLES

Create a Design Table to generate the series of Caster Frames and/or Caster Wheels as specified in the Table provided. See your instructor for specific instructions.

1. Generate an assembly drawing of the six **Caster Frames** produced from the Design Table and insert it into a drawing (see **Figure 6-30** as an example).

2. Generate an assembly drawing of the six **Caster Wheels** produced from the Design Table and insert it into a drawing (see **Figure 6-30** as an example).

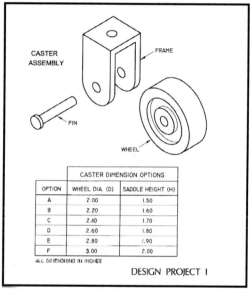

CASTER DIMENSION OPTIONS		
OPTION	WHEEL DIA. (D)	SADDLE HEIGHT (H)
A	2.00	1.50
B	2.20	1.60
C	2.40	1.70
D	2.60	1.80
E	2.80	1.90
F	3.00	2.00

ALL DIMENSIONS IN INCHES

DESIGN PROJECT I

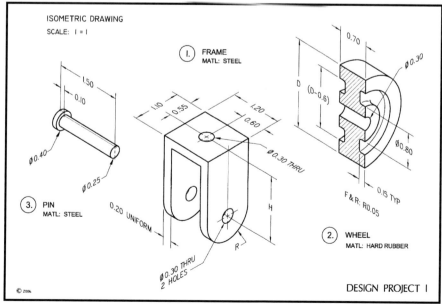

DESIGN PROJECT I

SUPPLEMENTARY EXERCISE 6-2: PULLEY AND PULLEY BRACKET DESIGN TABLES

Create a Design Table to generate the series of Caster Frames and/or Caster Wheels as specified in the Table provided. See your instructor for specific instructions.

1. Generate an assembly drawing of the six **Pulley Brackets** produced from the Design Table and insert it into a drawing (see **Figure 6-30** as an example).

2. Generate an assembly drawing of the six **Pulleys** produced from the Design Table and insert it into a drawing (see **Figure 6-30** as an example).

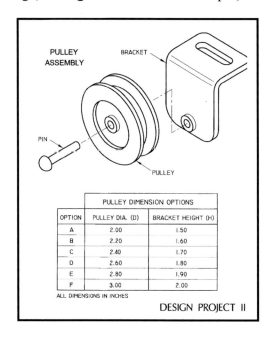

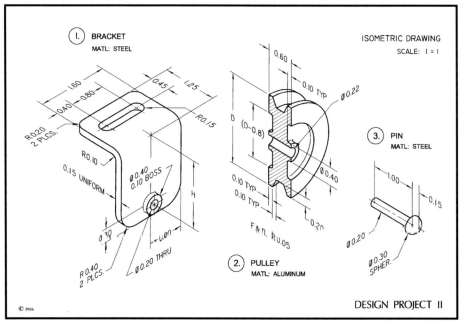

Computer Graphics Lab 7: Analysis and Design Modification II

INTRODUCTION

SOLIDWORKS parametric modeling software allows the designer to build and modify 3-D solid models easily by editing the dimensions and constraints imposed on the 2-D sketch, and then rebuilding the model. In this lab exercise, you will build a 3-D model of a pillow block model using an initial design concept. The initial 3-D model database will then be analyzed using the **SOLIDWORKS Simulation** add-in. You will perform a finite element analysis (FEA) on the pillow block to test the stress in the part during its primary function, which is to support a rotating shaft. The FEA results will then suggest modifications to the initial design of the pillow block in order to improve its strength and performance. You will then redesign the pillow block using the parametric features of SOLIDWORKS. Once the second design concept is completed, you will make a second study of the pillow block to see if the design changes improved the performance of the pillow block.

Procedure for Finite Element Analysis using SOLIDWORKS Simulation

The purpose of this exercise is to learn the stage in the design process subsequent to modeling the object. This is the Finite Element Analysis of the solid model. The results of this analysis are used to evaluate the design of the component. Finite Element Analysis allows the user to simulate a variety of load conditions that the part would be subjected to in the environment that it is designed for and evaluate the performance of the part under them before physically building the part. It is also extremely useful when destructive testing of the part in question is prohibitively expensive. SOLIDWORKS Simulation in conjunction with SOLIDWORKS offers a powerful tool for FEA. The part can be built as a solid model and SOLIDWORKS Simulation can be used to analyze the performance of the part.

The object of this exercise is to build a simple part (A Pillow Block) in SOLIDWORKS and analyze it using SOLIDWORKS Simulation. Before beginning the exercise SOLIDWORKS Simulation should be installed as an "Add In" to SOLIDWORKS. The steps for accomplishing this are outlined below:

1. Open SOLIDWORKS
2. Click on **TOOLS** - **Add-Ins** and select **SOLIDWORKS Simulation**

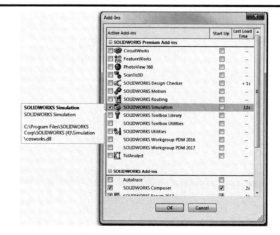

Figure 7-1. SOLIDWORKS Simulation Add-In.

This adds a Simulation Tab in the CommandManager. Right click in the gray area at the top of the screen and choose the Simulation Toolbar. The menu and the tools are displayed when a new file is created or an existing file is opened. The Simulation Tool Bar will have the following icons. The name of each of the icons is given in **Figure 7-2** and a written description of each of the commands is given in **Table 7-1**.

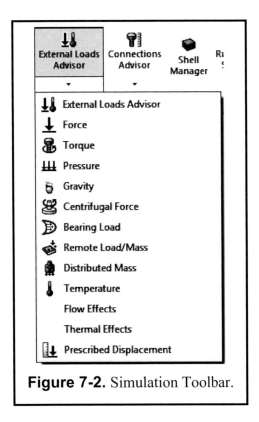

Figure 7-2. Simulation Toolbar.

External Loads Advisor
Force
Torque
Pressure
Gravity
Centrifugal Force
Bearing Load
Remote Load/Mass
Distributed Mass
Temperature
Flow Effects
Thermal Effects
Prescribed Displacement

Table 1. Definitions

Restraints	Applies restraints to the selected entities for the active structural study (static, frequency, or buckling).
Pressure	Applies pressure to the selected faces for the active structural study (static, frequency, or buckling).
Force	Applies force, torque, or moment to the selected entities for the active structural study (static, frequency, or buckling). The specified value is applied to each selected entity.
Gravity	Defines gravity loading for the active structural study (static, frequency, or buckling).
Centripetal Force	Applies centripetal forces for the active structural study (static, frequency, or buckling).
Remote Load	Applies remote loads for structural studies.
Rigid Connection	Applies rigid connection between selected faces in structural studies.
Bearing Load	Applies bearing loads on selected faces of different components.
Temperature	Applies temperatures on the selected entities for the active thermal or structural (static, frequency, or buckling) studies.

Exercise 7.1: FINITE ELEMENT ANALYSIS OF A PILLOW BLOCK

DESIGNING THE FIRST VERSION OF THE PILLOW BLOCK

You will start by building the preliminary Pillow Block. Study the needed geometry for this version in **Figure 7-3**. To start, open **ANSI-INCHES.prtdot** from your folder. Pull down **File** again, select **Save As**, key in the name **PILLOW1.sldprt**, make sure the directory is set to your folder, and then click the **Save** button. Notice that the part name is now "Pillow1" in the FeatureManager tree.

Select the **Front** plane in the FeatureManager. Add a new sketch by selecting the **Sketch** command on the Sketch Toolbar and draw the front profile with the center of the large hole at the **ORIGIN** and then extrude the profile using a **Mid-Plane** end condition to the depth of **1.40** inches (See **Figure 7-4**). In the **FeatureManager,** select the **Top Plane, add a new sketch for the holes and make an Extruded-Cut** according to the dimensions given in **Figure 7-5**.

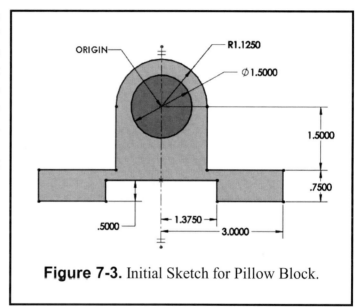

Figure 7-3. Initial Sketch for Pillow Block.

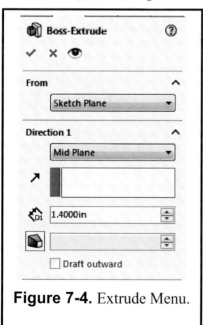

Figure 7-4. Extrude Menu.

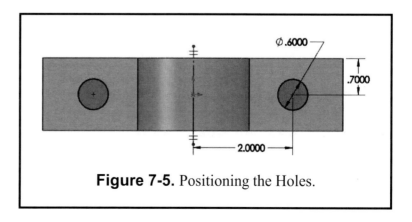

Figure 7-5. Positioning the Holes.

When finished, your solid model should look like the Pillow Block in **Figure 7-6**. **Save PILLOW1.sldprt** to your folder.

Now you will build a shaft that fits into the large hole of the Pillow Block. To start, open **ANSI-INCHES.prtdot**. Pull down **File** again, select **Save As**, type in the name **SHAFT.sldprt**, make sure the directory is set to your folder, and then click the **Save** button. Notice that the part name is now Shaft in the FeatureManager tree.

Select the **Front** plane in the FeatureManager. Click the **Sketch** command on the Sketch Toolbar and draw a **Circle** with a diameter of **1.50** inches at the origin (See **Figure 7-7**) and then extrude the profile about the **Mid-Plane** to the depth of **7.00** inches (See **Figure 7-8**). Add a **Chamfer** of **0.125** (**Figure 7-9**) on both ends of the shaft to complete this solid model. See **Figure 7-10**.

Save SHAFT.sldprt to your folder.

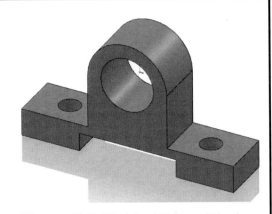

Figure 7-6. Finished Pillow Block.

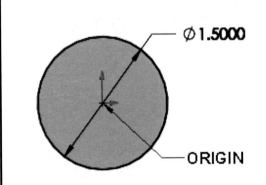

Figure 7-7. Circle Sketch for Shaft.

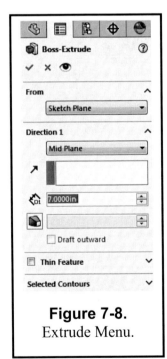

Figure 7-8. Extrude Menu.

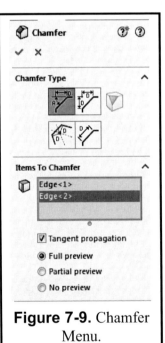

Figure 7-9. Chamfer Menu.

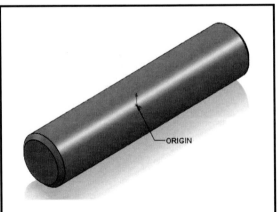

Figure 7-10. The Finished Shaft.

ASSEMBLY OF PARTS

At this time you are to make an assembly of the two parts. Make sure that you have Pillow1 and Shaft open. Go to **File**, **New** and select **Assembly**; set the **Units** to **MMGS**. Go to the pull-down **Tools** menu and select **Add-Ins**. Check **SOLIDWORKS Simulation** from the available list. Next, pull-down **View** and turn on **Origins**. **Save As "PILLOW ASSEMBLY-1.SLDASM."** Go to the **Window** pull-down menu and select "**Tile Vertically**." First, **Select** the Pillow Block and drag it to the **Origin (as your cursor approaches the origin, you will see a double set of arrows)** of the Assembly drawing. Next, **Select** the **Shaft** and drag it to the **Origin (as your cursor approaches the origin, you will see a double set of arrows)** of the Assembly Drawing. This will center the shaft inside of the hole of the pillow block. Maximize the Assembly file Window and view the assembly in isometric (see **Figure 7-11**).

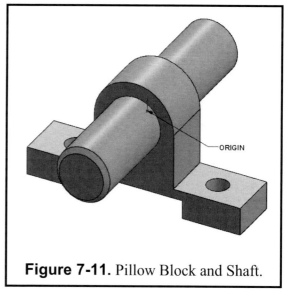

Figure 7-11. Pillow Block and Shaft.

FINITE ELEMENT ANALYSIS USING SOLIDWORKS SIMULATION

The actual analysis of the pillow block could be a complex procedure even with SOLIDWORKS Simulation, since analysis of the pillow block would have to include the effects of the rotating shaft and thermal stress on the shaft and the pillow block. In this case only a simple static analysis of the pillow block will be carried out.

If the Simulation Tab does not appear in the CommandManager, go to **Tools**, **Add-ins** and select **SOLIDWORKS Simulation** and select **OK**. This will place the pull down tab in the menu and the top of your screen.

To begin analyzing the part, a study needs to be defined. The steps required for defining a study are:

| File   Edit   View   Insert   Tools   Simulation   Window   Help   📌 |

Figure 7-12. Simulation Menu.

1. Click on the **SIMULATION TAB** at the top of the screen (**Figure 7-12**).
2. Select **New Study - New Study (Figure 7-13**).
3. Click in the box below **Study, Name**, and name the study (e.g. **Study-1- with your last name**) as shown in **Figure 7-14**, and select the type of study as **STATIC**. Then click **OK**.
4. The Study Manager Tree appears at the lower left of the screen. See **Figure 7-15**.

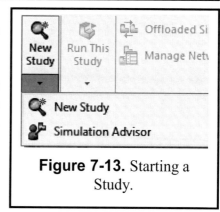

Figure 7-13. Starting a Study.

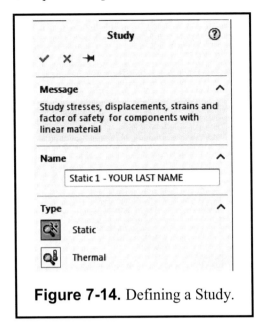

Figure 7-14. Defining a Study.

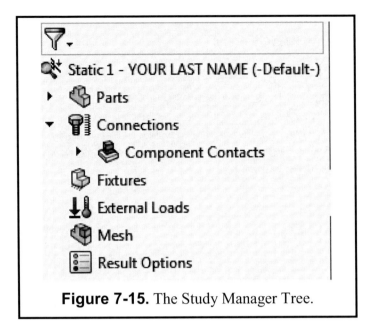

Figure 7-15. The Study Manager Tree.

Assigning Material to the Parts

SOLIDWORKS Simulation has a large material library, which contains the physical properties of a wide variety of materials. It also allows the user to create a new material with properties defined by the user. The procedure to apply a material is outlined below:

1. Go to the Study Manager Tree and click on the small arrow next to "Parts" to expand the **Parts** Feature. Right click on **Pillow1 (Figure 7-15)**, and select the **Apply/Edit Material** option.
2. In the Material window (**Figure 7-16**) select **From Library File – SOLIDWORKS materials.** Assign **Cast Alloy Steel** to the **Pillow Block.** Verify that the "**Type**" is set at **Linear Elastic Isotropic** and that the "**Units**" are set to **SI – N/mm^2 (MPa)** and click on **OK**. This assigns the material to the **Pillow Block.**
3. Repeat this process for the **Shaft** using **Alloy Steel** as the material.

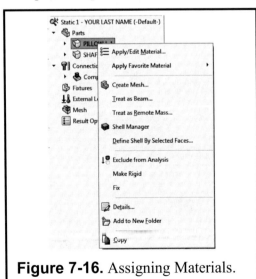

Figure 7-16. Assigning Materials.

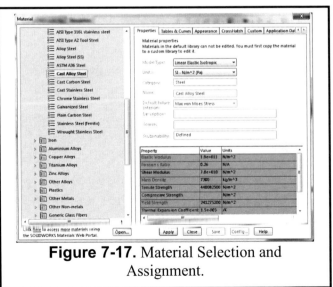

Figure 7-17. Material Selection and Assignment.

4. A *green check mark* appears on the icon next to the name of the model under the Solids folder in the SOLIDWORKS Simulation FeatureManager indicating that a material has been defined.

Applying Restraints:

A restraint is placed on the bottom surfaces of the pillow block to fix the object for a simple static analysis. The sequence of steps for this is follows:

1. Right click on **Fixtures** and select **Fixed Geometry.**
2. In the resulting window you are to select the two bottom faces of the Pillow Block as illustrated in **Figure 7-19**.

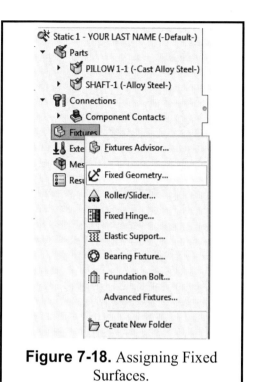

Figure 7-18. Assigning Fixed Surfaces.

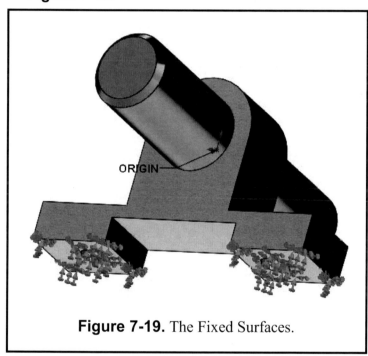

Figure 7-19. The Fixed Surfaces.

Applying the Force on the Shaft

In the static analysis the Pillow Block will react to the force the shaft exerts. The sequence of steps to apply this force is described below:

1. Right click on the **External Loads** icon in the SOLIDWORKS Simulation FeatureManager tree and select **Force** (see **Figure 7-20**).
2. For the **First Selection** box, click on the cylindrical surface of the shaft.
3. Next, choose **"Selected Direction."**

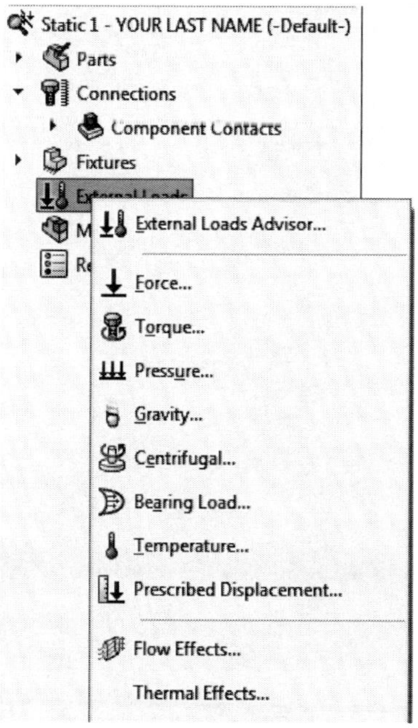

Figure 7-20. Applying External Loads.

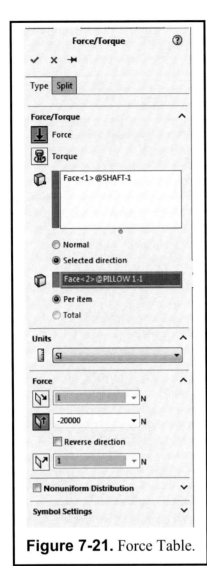

Figure 7-21. Force Table.

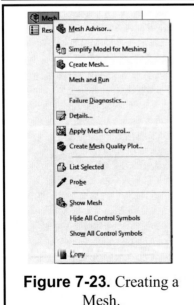

Figure 7-23. Creating a Mesh.

4. For the Second Selection box, pick the **Front Vertical surface of the Pillow Block**.
5. Set the units to the **SI** system.
6. Under **Force,** activate the second force and enter the value of the force **(-20000 N)** and click **OK** (**Figure 7-21**). Forces appear as lavender arrows. _Note: Make sure that the lavender arrows are pointing down._

This completes the application of restraints and forces (**Figure 7-22**) to the Pillow Block and the Shaft. The next step is to create the Mesh required for analysis.

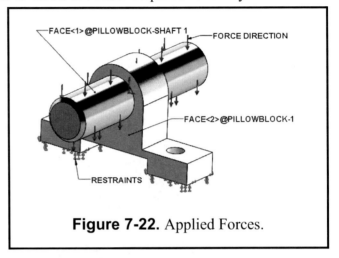

Figure 7-22. Applied Forces.

Creating the Mesh:

The mesh is the division of the solid into a number of small geometric divisions. The effect of the stress on these divisions is computed and then integrated to provide a stress analysis of the entire part. This is an extremely simplistic way of looking at the mesh generation process, which in reality is extremely complicated and takes a relatively long time to execute even on today's fastest computers. The procedure to create the mesh for this analysis is detailed below:

In the SOLIDWORKS Simulation Manager tree, right click on **Mesh** and select **Create Mesh** (**Figure 7-23**).

1. This opens up the mesh dialog box, which allows the user to define the mesh size for the analysis. A smaller mesh yields better results but takes longer to mesh as well as analyze. Hence it is a tradeoff between mesh size and computation time. Change the units to **mm**. Set the **Global Size** to **5** for a slightly finer mesh and **Tolerance** to **.25 (See Figure 7-24)**.

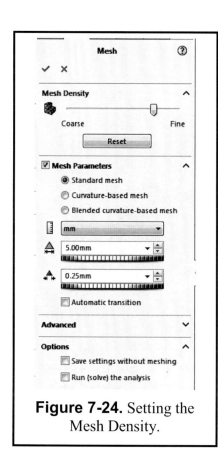

Figure 7-24. Setting the Mesh Density.

2. Click on **OK** to accept the values and the Meshing process begins.

3. At the end of the meshing process, a check mark is placed against Mesh in the SOLIDWORKS Simulation FeatureManager Tree. This causes the mesh to be displayed on the solid as shown in **Figure 7-25**.

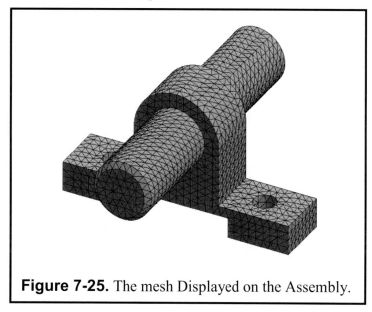

Figure 7-25. The mesh Displayed on the Assembly.

To Run Static Analysis

The mesh generation is now complete and all parameters required for running the analysis have been defined. To run the Static Analysis, right click on the **Name of the Study** and select **Run**. The analysis begins. This process also takes a long time to complete and the execution time depends on the mesh size. The program displays a message once the static analysis is completed. Click on **OK**.

Post-processing

In this section the analysis has been completed and the results are visualized and interpreted. If the study ran successfully, a new item (**RESULTS**) will appear in the Manager tree. Expand the results folder. Three new items now appear in the **SOLIDWORKS Simulation** FeatureManager (**Figure 7-26**). They are **Stress, Displacement,** and **Strain.** You can right-click on any of these items and select the action you wish to take (see **Figure 7-27**). **SHOW** illustrates what the study has computed, **DELETE** is obvious, and **COPY** puts the results on the computer clipboard. Right Click on **STRESS1** and select **Show**. You should get a figure that resembles **Figure 7-28**.

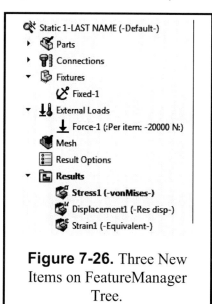

Figure 7-26. Three New Items on FeatureManager Tree.

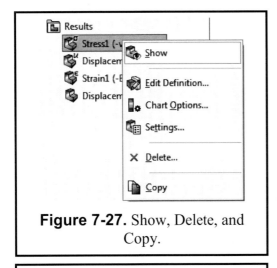

Figure 7-27. Show, Delete, and Copy.

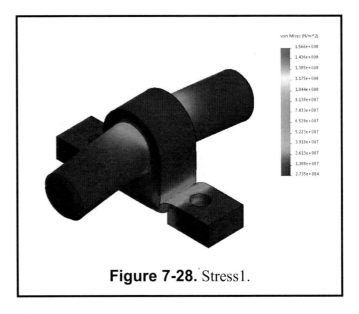

Figure 7-28. Stress1.

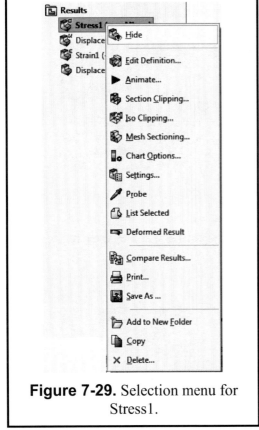

Figure 7-29. Selection menu for Stress1.

Right-click on **STRESS1** a second time to get a selection window as shown in **Figure 7-29**. Select several of the options to see what results are generated. If you select the **Edit Definition** you will notice that the Scale Factor is approximately 357. The distortion of the image on the computer is exaggerated by a factor of 357. If you select **Animate** the image will be put into motion. **Section Clipping** lets you manipulate a cross section to see internal forces and distortions.

Go to the **Stress1** Selection menu and choose **Print.** Go to **Properties** and change the print format to **Landscape**. Then click on **OK** and the plot will be printed out.

Insert the **Pillow-1 Assembly** rendered image onto a **Title Block** drawing and **Print** it. This sheet along with the print of the **Stress1 - Von Mises** are to be turned in to your instructor.

ANALYZING THE PILLOW BLOCK FOR STRESSES

One of the powerful applications of solid modeling using SOLIDWORKS is the ability to easily interface the solid models with analysis software. In this case, the Pillow Block was analyzed for stress distributions when forces were applied to it during its normal function, which is to support a rotating shaft.

The FEA study determines how the stresses and strains were distributed throughout the solid material of the Pillow Block when the given forces were applied. The results are readily displayed as color contour maps that show the stress distribution. Areas where stresses concentrate can be detected and the FEA results can lead to an improved design of the part. **Figure 7-28** shows the results of this FEA study. It can be seen that the stresses tend to concentrate in the thin cylindrical wall surrounding the shaft hole, and also at the right angle where the upright feature emerges from the Pillow Block base. These two design features are likely candidates for further consideration and re-design when you create the next version of the Pillow Block. With the results of all this analysis, you are now ready to start the design modification stage.

DESIGN MODIFICATION OF THE PILLOW BLOCK

After studying the results of the FEA, several modifications are suggested for the pillow block to create a second version.

EDITING THE PILLOW BLOCK

Make sure the former **Pillow1** part file is **OPEN** on your computer screen, and then **SAVE AS Pillow2** to start a new version of the part. You will now see how easy it is to make most of the design modifications by simply editing the sketch that was used to extrude the base part. In the FeatureManager tree, expand the **Base/Extrude** feature to reveal **Sketch1**. Select **Sketch1** and from the context toolbar select the **Edit Sketch** option. Using either **Table 2** "Design Modification Table – A" and/or **Figure 7-30**, make the changes to modify the sketch to conform to the new criteria. When this is completed repeat the above process on the **Cut-Extrude** feature to relocate the holes using **Table 2** "Design Modification Table – A" and/or **Figure 7-31**.

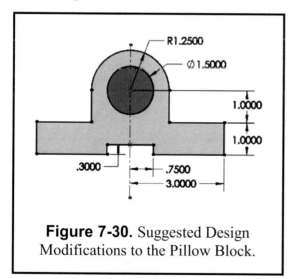

Figure 7-30. Suggested Design Modifications to the Pillow Block.

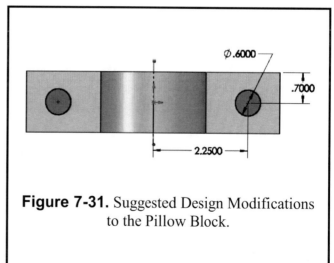

Figure 7-31. Suggested Design Modifications to the Pillow Block.

Table 2. Design Modification Table – A.	
Feature	**Change Order Request**
Outer Radius Around Shaft Hole	Change Radius from 1.125 to 1.25 inches
Change the thickness of the Base	Change Height from .75 inches to 1.00 inch
Center of the Shaft Hole Distance from Base	Change Height from 1.50 to 1.00 inch
Vertical Bolt Holes	Change Position from 2.00 to 2.25 inches from the centerline

Basically, the thickness of the material around the shaft hole was too thin and the outer radius of the pillow block needs to be increased in order to thicken the material surrounding the hole. Sharp 90 degree angles are generally not a good idea for mechanical design, and fillets can be added to bolster the strength at these junctures. Also, the slot, which is used primarily to save on material waste and part weight, is poorly designed. The width should be reduced and fillets added to compensate for the downward pressure. Finally the vertical bolt holes (**Figure 7-31**) need to be centered on the new foot-pad bases.

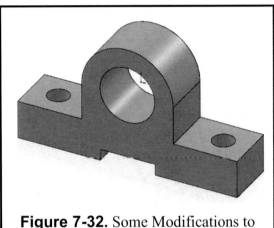

Figure 7-32. Some Modifications to the Pillow Block.

Once all the current changes have been made, select **Rebuild**. The new Pillow2 will appear in its current state (**Figure 7-32**).

After the design has been updated to look like **Figure 7-32**, **add fillets** to the Pillow block according to **Table 3** –"**Design Modification Table - B**." When this is completed, your model should look like the Pillow Block in **Figure 7-33**.

Table 3. Design Modification Table - B	
Feature	**Change Order Request**
Right Angle Surface Intersections (2 Places)	Add Fillets with .375 inch Radius
Bottom Slot – Upper Two Corners	Add .25 inch Fillets
The Upper Edges of the Base	Add .25 inch Fillets

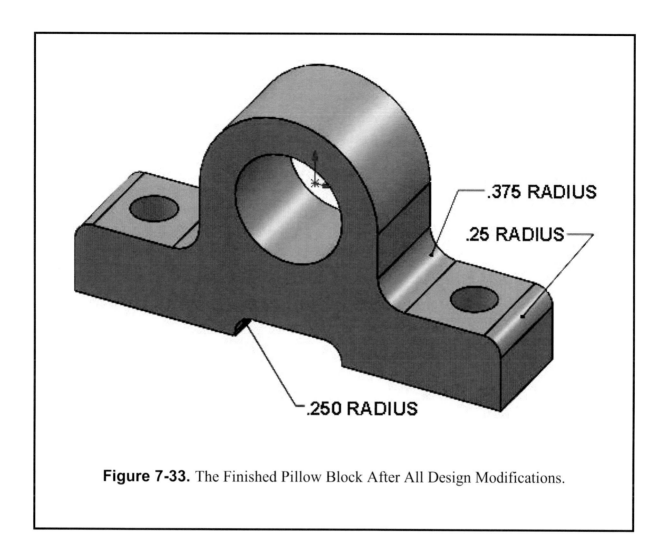

.375 RADIUS

.25 RADIUS

.250 RADIUS

Figure 7-33. The Finished Pillow Block After All Design Modifications.

Study – 2 – LAST NAME

After you have finished the design modifications, go to **File**, select **New**. Click on **Assembly** and **Save As "Pillow Assembly-2."** Repeat the assembly process you used to create the first assembly. The two parts that make up the new assembly are **Pillow-2** and the **Shaft**.

You will now **Repeat** the **SOLIDWORKS Simulation** process, using the same criteria and procedure performed on the original design. **Save** the **Assembly** and the **SOLIDWORKS Simulation/Study2 – Your last name** to your folder. Insert the **Pillow-2 Assembly** rendered image onto a **Title Block** drawing and **Print** it. This sheet along with the prints of the **Stress1 - Von Mises** for the second study are to be turned in to your instructor.

Exercise 7.2: Thermal Analysis of a Computer Chip

A computer chip is one of the many components in a computer, yet it contributes significantly to the very thing that may damage its performance. Computer designers must consider temperature, fatigue, and chemical and material properties among the many factors that may affect the performance. Heat has to be a major consideration when designing a computer system. As heat builds up in a component, the function and integrity of that component and even the entire system may be affected.

In this exercise you will construct a 3D solid model of a computer chip and a heat sink to see how one affects the other. Then the material properties, loads, reactions, and constraints typical for such a part will be presented. Finite element analysis will be used to visualize the resulting distributed heat through the chip and then with the heat sink attached. The first portion of this exercise will be to do a steady state study. You will then use the results to do a Transient thermal analysis.

CONSTRUCTING THE HEAT SINK

We begin with a sketch for the Heat Sink. Open **ANSI-INCHES.prtdot** from your folder. Do a **Save As – Heat Sink.sldprt**. Under the **Tools – Options – Document Properties** select **Units** and change them to **7 decimal places**. Select **Front** as the sketch plane, go to **Sketch** mode and draw a **Vertical Centerline** to the left of the origin. Go to **TOOLS - SKETCH TOOLS** and **SELECT – DYNAMIC MIRROR**. Now **SELECT** the **LINE** command and draw an angled line downward and another very short horizontal line to the centerline. Dimension the lines according to the sketch in **Figure 7-34**. Add a Relation so that the end points of the

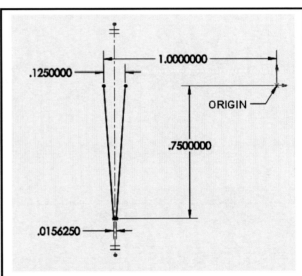

Figure 7-34. Base Drawing of the Fin for the Heat Sink.

two angled lines are horizontal with the Origin. A detail of the very small end of the pattern is shown in **Figure 7-35**. When finished, select the **Linear Sketch Pattern**. Set **Distance-1** to **.3125** and the **Instance Count** to **7**. In the box labeled **Entities to Pattern** at the bottom of the Linear Pattern display, **Select** the two **Diagonal Lines** and the very **Short Bottom line** to be patterned. The pattern should extend over the Origin. The completed model will have the Origin on the top surface and in the center of the model. This will help in the assembly of the parts for analysis. The next step is to

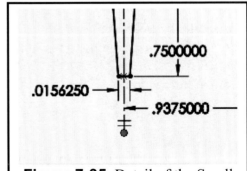

Figure 7-35. Detail of the Small End of the Fin Profile.

draw a **Horizontal Line** all the way across the top of the pattern. Then **Offset** that horizontal line by **.125** inches downward. You will have to un-select "Select Chain" for this to complete. Your resulting figure should look like **Figure 7-36**.

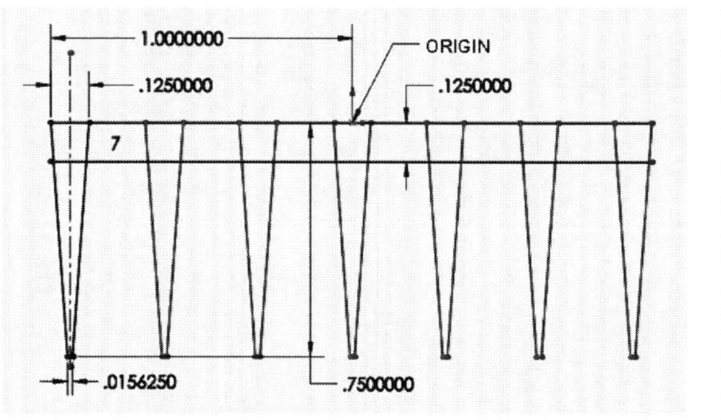

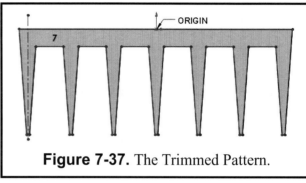

Figure 7-36. The Two Horizontal Lines.

Next, **Trim** the **angled lines** above the **offset line** and the **offset lines** between the **angled lines**. Be sure to also **Trim** the two very short segments of the offset line that extend beyond the outer edges of the pattern. Your completed sketch should appear as in **Figure 7-37**.

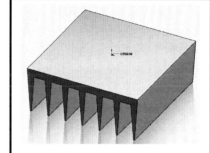

Figure 7-37. The Trimmed Pattern.

To complete the Heat Sink, **Extrude** the Sketch – **Mid-Plane** for **2** inches. Your solid model should look like the model in **Figure 7-38**. You will now assign the material to the Heat Sink. **Right click** on **Material <not specified>** in the FeatureManager Tree, **Edit Material**, expand **Aluminum Alloys** and select **1060 Alloy**. Also activate the **Appearance Tab** to make sure that the "**Apply Appearance of: 1060 Alloy**" is **unchecked**. Click **Apply** toward the bottom of the window and then **Close** your Material Window. You have now completed the model of the Heat Sink. You may assign a color to the Heat Sink. Save your model in your File Folder as **Heat Sink.sldprt**.

CONSTRUCTING THE CERAMIC COMPUTER CHIP

Begin with a sketch for the Ceramic Computer Chip. Open **ANSI-INCHES.prtdot** from your folder. Do a **Save As**

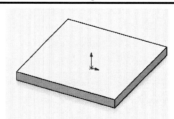

Figure 7-38. The Finished Heat Sink.

CERAMIC COMPUTER CHIP.sldprt. Select **Top** as the sketch plane, go to **Sketch** mode, and create the profile as shown in **Figure 7-39**. This is a simple square that measures **1.50** inches centered at the **Origin**. Click on the Features Tab and **Extrude** the Sketch **Upward .125**

Inches. **Right click** on **Material <not specified>** and **Edit Material**. Expand **Other Non-Metals** and Select **Ceramic Porcelain**. Under the **Appearance Tab** make sure the **Apply Appearance** box is unchecked. You may assign a color to the Computer Chip. Save your model in your File Folder as **Computer Chip.sldprt**.

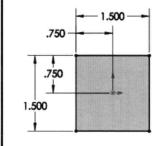

Figure 7-39. Sketch of Computer Chip.

Figure 7-40. Completed Ceramic Computer Chip.

CONSTRUCTING THE ASSEMBLY

With the Heat sink and Ceramic Computer Chip parts open, go to **File – New – Assembly**. In the FeatureManager Tree select the **Heat Sink** and **Okay** to Begin Assembly. This will place the Heat sink at the Origin. Next, **Insert** the **Ceramic Computer Chip** at the **Origin**. This will automatically mate the two solids to finish the assembly. **Save** the assembly to your file folder as **Thermal Study.sldasm**. Your assembly should look like **Figure 7-41**.

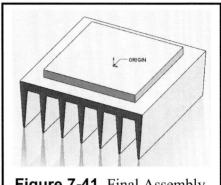

Figure 7-41. Final Assembly.

THERMAL ANALYSIS OF THE CERAMIC COMPUTER CHIP

Starting the Study

Go to the **Tools** tab – **Add-Ins** and select **SOLIDWORKS Simulation** and click **OK**. This will insert a pull down tab named **Simulation**. Activate the Simulation tab. Under **Study Advisor** select **New Study**. In the box labeled **Name**, enter **Thermal Study 1 - (your last name)**. In the **Type** box activate the **Thermal** Icon as shown in **Figure 7-42. Okay** your settings.

In the Thermal **Study** manager tree, right click on **Connections** and select **Contact Set**. Since you only have two pieces and they are in contact with each other, you can **Click** on "**Automatically find Contact Sets**." Under Options **Select Touching Faces**. Under Components **Select** the **Two Parts** of the Assembly. Under Results make sure that the **Type** is set to "**Thermal Resistance**." Under Thermal Resistance **Select - Distributed** and insert the **SI** units of **.0001 K-m^2/W**. Under the **Advanced** box select **Surface to Surface**. See **Figure 7-43** for the settings. **Okay** all of the settings.

Figure 7-42. Thermal Study Set-up.

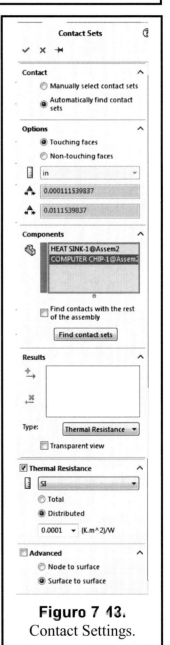

Figuro 7 13. Contact Settings.

Identifying Heat Source and Convection Settings

The heat produced in a computer chip will vary depending on its complexity. For this exercise, 15 Watts of power will be used. The next phase of this study is to identify the amount of heat and what is creating the heat. **Right** click on **Thermal Loads** and select **Heat Power**. This opens a panel in the FeatureManager area for input. **Right Click** on the **Heat Sink** and **Hide** it so you can rotate the **Computer Chip** to **Select** the **Bottom** surface. This is the surface that will be transferring the heat to the Heat Sink. Your Surface selection will appear in the light blue box. Under the **Heat Power** enter **15 Watts**. When completed **Okay** the selections. See **Figure 7-44** for the settings. Before you proceed to the next step, **Right Click** on the **Heat Sink** and **Click** on "**Show.**"

The last step before the mesh is created is to identify the surfaces that will disburse the heat from the Computer Chip. **Right Click** on **Thermal Loads** and select **Convection**. Now **Select** all of the surfaces of the **Heat Sink except** for the **Top Surface (there are 29 Surfaces that should be selected**). Now set the **Convection Coefficient** to **15 W/(m^2.K)** and set the **Bulk Ambient Temperature** to **300** degrees Kelvin. These settings can be seen in **Figure 7-45**.

Creating the Mesh

The Mesh is the method of dividing the solid model into an ordered set of finite geometric elements that can be mathematically analyzed to determine the effects of stress, strain or temperature on a solid.

Right Click on **Mesh** in the FeatureManager Tree and select **Create Mesh**. Under the **Mesh Parameters** change the **Units** to **mm**, the **Global size** to **2.00mm** and the **Tolerance** to **0.10mm** as shown in **Figure 7-46**. When you click OK, the program will Mesh the assembly as shown in **Figure 7-47**.

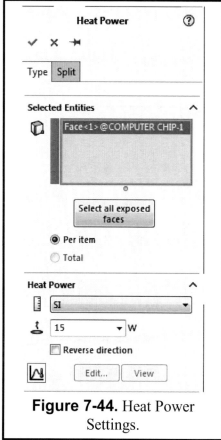

Figure 7-44. Heat Power Settings.

Figure 7-45. Convection Settings.

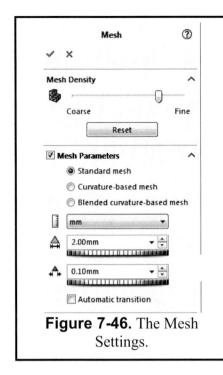

Figure 7-46. The Mesh Settings.

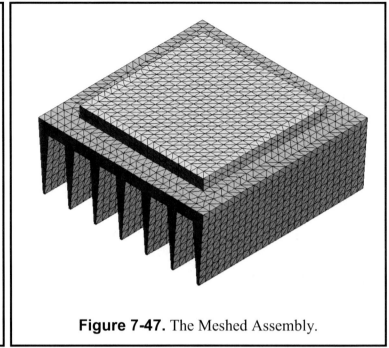

Figure 7-47. The Meshed Assembly.

Running the Analysis

Right Click on the name of the study in the **SOLIDWORKS Simulation** FeatureManager and select **Run**. There will be some delay (up to a Minute) in getting the results of the study because of the amount of calculations needed to complete the study. The FeatureManager Tree will now display a new listing identified as **Results: Thermal1 (-Temperature-)** along with an image of the results of the study (See **Figure 7-48**).

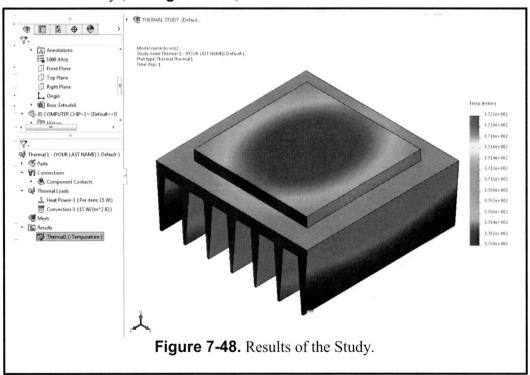

Figure 7-48. Results of the Study.

Right click on **Thermal1 (-Temperature-)** and select **Print** to get a hard copy of your Thermal Study.

The next step is to **Right Click** on **Thermal1 (-Temperature-)** and select **Edit Definition**. Under **Display** activate the Down Arrow of the **TEMP: Temperature** box and **Select** the last option – **HFLUXN: Resultant Heat Flux, Figure 7-49**. Then under **Advanced Options**, check the box next to **Show as vector plot**.

It is possible to adjust the density and size of the Vector Plot. **Right Click** on **Thermal1 (-Res heat flux-)** and **Select - Vector Plot Options**. Enter **500** in the first box and **25** in the second box (**See Figure 7-50**). You will then get a result that shows the direction of heat flow (**Figure 7-51**). There are arrows going straight down out of the Computer Chip and disbursing out through the fins of the Heat Sink. **Right Click** on **Thermal1** and **Print** a copy of the Heat Disbursement Illustration.

Insert your **Assembly** on a **Title sheet** with the usual annotations and submit this along with the prints of **Thermal1 (-Temperature-)** and **Thermal1 Vector Plot** to your instructor.

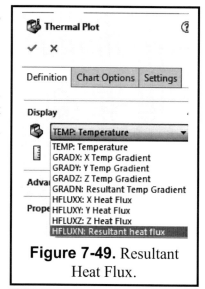

Figure 7-49. Resultant Heat Flux.

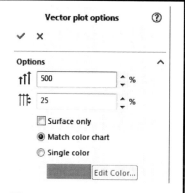

Figure 7-50. Vector Plot Options.

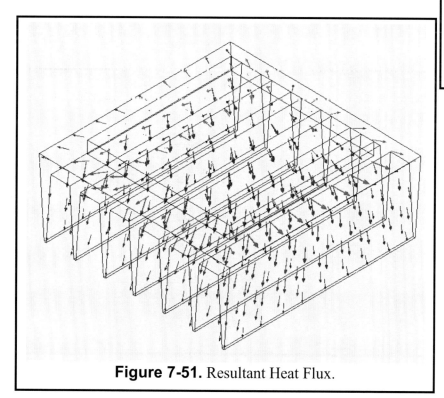

Figure 7-51. Resultant Heat Flux.

NOTES:

Computer Graphics Lab 8: Kinematics Animation and Rapid Prototyping

In Computer Graphics Lab 8, you will explode a previously built assembly file using SOLIDWORKS "Assembly Exploder." You will set the direction and distance of the exploded parts of the assembly. You will then create a simple animation of this exploding assembly using SOLIDWORKS "Animation Wizard." You will save the animation as an .AVI file that can be played on an external viewer, like Windows Media Player. You will also be briefly introduced to SOLIDWORKS Physical Simulation capabilities. A second objective of this Computer Graphics Lab 8 is to produce an .STL (stereo lithography) file of your SOLIDWORKS parts. The .STL files are then transferred to a rapid prototyping machine to make physical prototypes of the parts.

ASSEMBLY TOOLBAR

In the earlier Computer Lab 5, you were introduced to the **Assembly Toolbar** as shown in **Figure 8-1**. When you open a completed assembly file, you can use these tools to manipulate the assembly components. When you click on the **Exploded View** command, the **Explode** properties are displayed, as shown in **Figure 8-2**. This is where you can define the options for each of the explosion steps of the assembly.

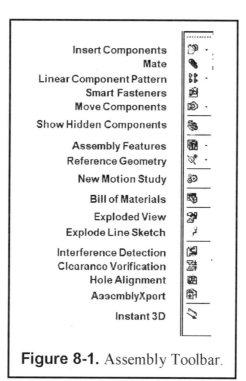

Insert Components
Mate
Linear Component Pattern
Smart Fasteners
Move Components
Show Hidden Components
Assembly Features
Reference Geometry
New Motion Study
Bill of Materials
Exploded View
Explode Line Sketch
Interference Detection
Clearance Verification
Hole Alignment
AssemblyXport
Instant 3D

Figure 8-1. Assembly Toolbar.

Figure 8-2. Explode View Options.

For example, in **Figure 8-3** the **Pin** of the Terminal Support assembly has been exploded upward as shown. The user first would activate the **Settings** selection box and then select the **Pin** on the assembly. The **Pin** is then listed in the selection box of the components to be exploded. Three direction arrows appear on the selected component. Select the arrow in the direction that you want to explode the part, and then type the **Distance** to explode; make it **6.00** inches. If needed, toggle the **Reverse Direction** button so the Pin goes upward. To complete the exploded step, select the **Add Step** button. *Explode Step 1* is listed in the **Explode Steps** list at the top.

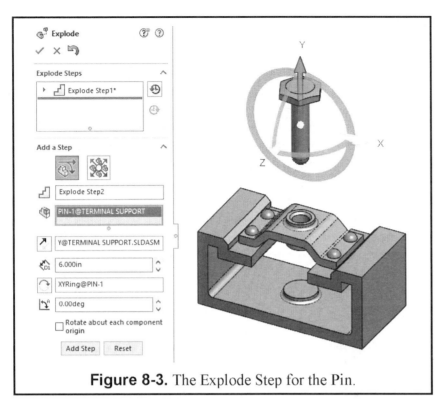

Figure 8-3. The Explode Step for the Pin.

ANIMATION WIZARD

The "Animation Wizard" is an easy tool to create simple animations when you have a part or assembly file loaded. From the Assembly tab select **New Motion Study** (See **Figure 8-4**).

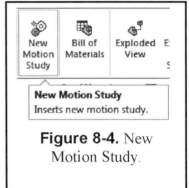

New Motion Study
Inserts new motion study.

Figure 8-4. New Motion Study.

An animation Controller window appears in the lower pane of the screen. Select the Animation Wizard icon (See **Figure 8-5**). It has three options (See **Figure 8-6**):

> Rotate Model
> Explode Model
> Collapse Model

If you select **Rotate Model**, it asks for an axis to rotate around and the number of rotations. You then get to the next step (see **Figure 8-7**), which asks about the duration (in seconds) for the animation.

In order to select the other two options in the "Animation Wizard" you must open an assembly which already has an **Exploded View**. You can then create an **Explode** animation or a **Collapse** (reverse of explode) animation.

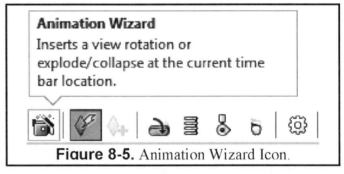

Animation Wizard
Inserts a view rotation or explode/collapse at the current time bar location.

Figure 8-5. Animation Wizard Icon.

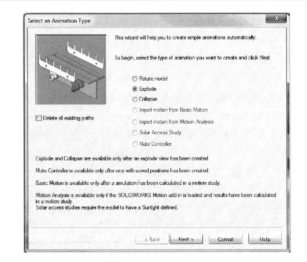

Figure 8-6. The Animation Wizard Create Menu.

ANIMATION CONTROLLER AND MOTION MANAGER

The "Animation Controller and MotionManager" appears on the bottom of the screen **(Figure 8-8)** when you select a **New Motion Study** as illustrated in **Figure 8-4**.

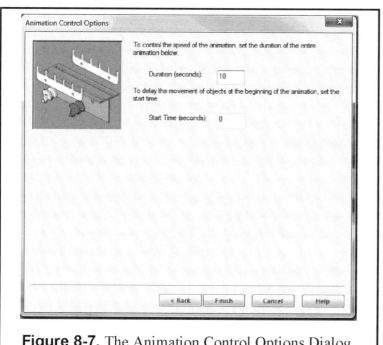

Figure 8-7. The Animation Control Options Dialog.

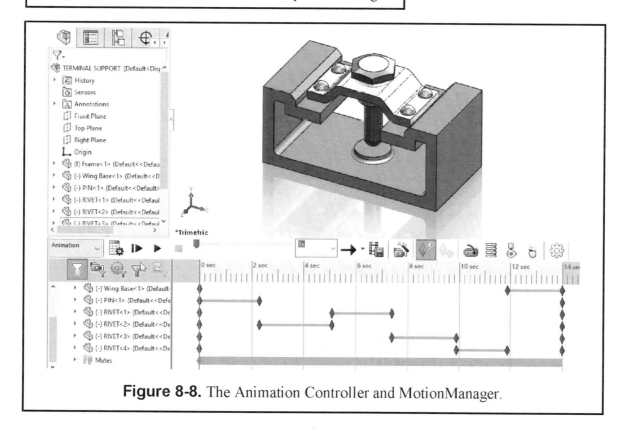

Figure 8-8. The Animation Controller and MotionManager.

The following button commands are available on the "Animation Controller" (**Figure 8-9**).

Play from Start button stops the animation and starts to play from the beginning.

Play button plays the animation at the specified speed.

Stop button stops the animation and sends it back to the start position.

There are three "**Playback Modes**" on the controller.

Figure 8-9. The Animation Controller.

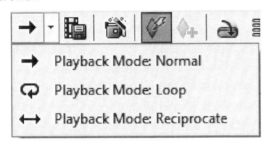

Normal Mode button sets the animation to run once from start to end at the specified speed.

Loop Mode button sets the animation to run over and over again in a continuous loop.

Reciprocate Mode button sets the animation to first run forward, and then run backwards, and repeat this reciprocation in a continuous fashion.

Save as AVI File button will open the "Save Animation to File" menu and allows you to save the animation as an .AVI file. It can then be played on an external media player.

Animation Wizard launches the "Animation Wizard" tool (see **Figures 8-4 and 8-5**).

Zoom In and **Zoom Out** buttons allow you to zoom in the MotionManager to see smaller or larger time intervals.

INTRODUCTION TO PHYSICAL SIMULATION

Physical Simulation allows you to simulate the effects of motors, springs, and gravity on your assemblies. Physical Simulation combines simulation elements with SOLIDWORKS tools such as mates and Physical Dynamics to move components around your assembly. These icons appear on the top of the MotionManager area.

The options on the Simulation Toolbar are shown in Figure 8-10, and include the ability to add a Linear Motor, Rotary Motor, a Linear Spring, or Gravity to your model.

Figure 8-10. The Simulation Toolbar.

Exercise 8.1: Exploded Animation of the TERMINAL SUPPORT ASSEMBLY

Recall in Computer Graphics Lab 5 you created some assembly models in SOLIDWORKS. If you created and saved the Terminal Support Assembly, you can now work this Exercise 8.1. If you created and saved the Pulley Assembly, you should go to Exercise 8.2. If you did not save either assembly file, then you need to return to Computer Graphics Lab 5 and re-build one of the assemblies first and then return to this Lab 8.

CREATE EXPLODED ASSEMBLY VIEW

Open the previously made **Terminal Support.sldasm** assembly from your designated folder. Before you can create the animation, first you need to add an Exploded View to the assembly. To do this, choose the **Insert** menu and click on **Exploded View** (or click on the Exploded View command from the Assembly Toolbar). The "Explode" options are now shown, as in the earlier **Figure 8-2.**

The first part to explode is the **Pin**. Activate the "Components to Explode" selection box in the **Settings** section. Next, select the **Pin** in the assembly, and it is added to the Components to Explode selection list. Multiple components can be selected at the same time.

Now, from the coordinate triad added to the component, select the **Y direction arrow** to define the explode direction. The explode direction will be added to the Explode Direction selection box, in this case *Y@terminal support*.

Then enter an explode **Distance** of **8.00** inches. If needed, toggle the **Reverse Direction** button so the Pin goes upward. To complete the explode step click on the **Add Step** button, and *Explode Step 1* is added to the Explode Steps list at the top. The correct explode settings for the **Pin** are shown in **Figure 8-11**.

Figure 8-11. The Explode Step Settings for the Pin.

Now explode the four Rivets either individually or together. The Explode Step Components selection box is automatically activated. Select the four Rivets. They are added to the selection list. Go to the assembly and select the **Y direction arrow** to define the explode direction. The explode direction will be added to the Explode Direction selection box, in this case *Y@terminal support*. Then enter a **Distance** of **7.00** inches. If needed, toggle the **Reverse Direction** button so the Rivets go upward. To complete the explode step click on the **Add Step** button, and *Explode Step 2* is added to the Explode Steps list at the top.

The correct explode settings for the Rivets are shown in **Figure 8-12**.

Figure 8-12. "Explode" Settings for the Rivets.

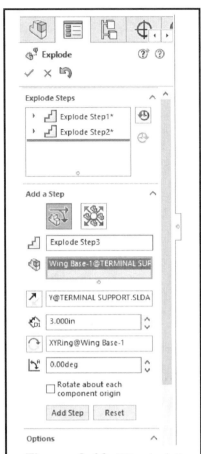

Figure 8-13. "Explode" Settings for the Wing.

Finally, explode the Wing Base part. Select the Wing Base from the assembly. It is added to the selection list. Go to the assembly and select the **Y direction arrow** to define the explode direction. The explode direction will be added to the Explode Direction selection box, in this case *Y@terminal support*. Then enter a **Distance** of **3.00** inches. If needed, toggle the **Reverse Direction** button so the Pin goes upward. To complete the explode step click on the **Add Step** button, and *Explode Step 3* is added to the Explode Steps list at the top.

The correct explode settings for the Wing Base are shown in **Figure 8-13**. The Terminal Support is now exploded, as shown in **Figure 8-14**.

After the three exploded steps are added to the assembly, click **OK** to finish the **Explode** command.

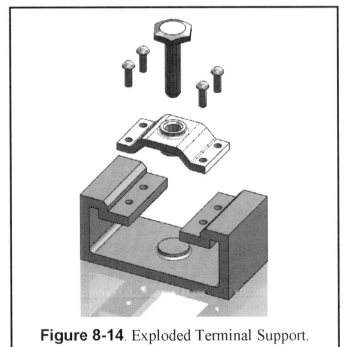

Figure 8-14. Exploded Terminal Support.

CREATING THE ANIMATION

Now you will create the animation using the Animator Wizard. Click on **NEW MOTION STUDY** in the Assembly tab of the CommandManager. The "Animator Controller and MotionManager" area appears on the bottom of the screen as shown in earlier **Figure 8-8**. Select the **Animation Wizard** 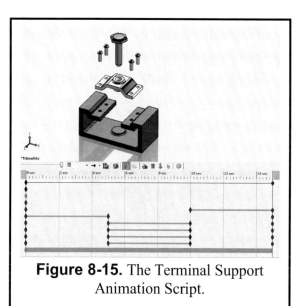 command from the animation toolbar (see the choices on **Figure 8-6**). Select the **Explode** option on the "Select an Animation Type" dialog. Then click the **Next** button. In the next dialog (see **Figure 8-7**) enter the animation "Duration" of **15** seconds and "Start Time" as **0**. Finally, click the **Finish** button.

PLAYING THE ANIMATION

In the "Animation Controller" shown in **Figure 8-8,** press the **Play** button to see the animation on your screen. Try some of the other animation options. **Play** the animation in the **Normal Mode. Play** the animation in the **Loop Mode** and see the assembly move in a repeated fashion until you click **Stop**. **Play** the animation in the **Reciprocate Mode** and see the assembly move in a reciprocating mode until you click **Stop**. Review the different combinations to play the exploded animation (see **Figure 8-15**). When you have selected **Normal Mode**, <u>show it to your instructor or lab teaching assistant.</u>

Figure 8-15. The Terminal Support Animation Script.

SAVING THE ANIMATION

Return to the "Animation Controller" buttons and select the **Save Animation** command (). The "Save Animation to File" dialog now appears on the screen as shown in **Figure 8-16**. Name the file **Terminal Support** and use an **.AVI file** type. Use the **SOLIDWORKS screen** as the Renderer and set the animation rate to **15** frames per second. Then click the **Save** button. Click on **OK** in the "Video Compression" dialog box that next appears, in order to accept the default settings and save the animation. This causes the exploded model animation to be saved to a file as well as played on screen. It may move slowly at this point because it is calculating the motion and saving the video file. <u>Note:</u> You may have to adjust your model viewport to fit the whole animation on your screen as it is being captured.

Figure 8-16. "Save Animation" Menu.

At this point you are finished with the animation, so you may want to **Save** your exploded assembly file as **EXPLODED TERMINAL SUPPORT**.

PHYSICAL SIMULATION EXAMPLE

Open the original **Terminal Support** assembly file from your designated folder. The assembly components should be completely mated in position. Now you need to remove one of the mates. Expand the **Mates** folder in the FeatureManager Tree, and review all the mates in the **Terminal Support** assembly. Right click on the **Coincident(x) Wing Base<1>,Pin<1>** mate, and select **Delete** as shown in **Figure 8-17**. Answer **Yes** in the **Delete** confirmation dialog. Deleting this mate allows you to move the Pin upward, but still concentric to the hole.

Select the **Move Component** command on the Assembly Toolbar, or simply click and drag the Pin to move it up from its mated position above the assembly. Click on the **New Motion Study** command. Next select the **Motor** command ⌖ from the Simulation Toolbar (see **Figure 8-10**). Using the **Rotary Motor** option, select the cylindrical shaft of the Pin. Next, set the **RPM** below the Constant Speed to **25 RPM** as shown in **Figure 8-18**. Then click **OK** to add the motor. Now select the **Calculate Simulation** button on the Simulation Toolbar. The computer will now calculate the physical simulation and play it on the screen as shown in **Figure 8-19**. **If instructed**, <u>show it to your instructor</u> and then press the **Stop** button on the Simulation Toolbar.

You can now click the **Replay Simulation** button on the Simulation Toolbar if you want. Now try to add a **Linear Motor** to one of the Rivets. You will have to **Delete** a **Coincident** mate holding the Rivet in place to get the Rivet to move away from the Wing Base. Then click the **Calculate Simulation** command to start the new simulation.

Before you leave the lab, your instructor needs a hard copy of your work. **Open** your previously saved assembly file, **Explode** it, and **Print** an isometric shaded view of the exploded assembly on a Title Block drawing sheet, as shown in **Figure 8-20**.

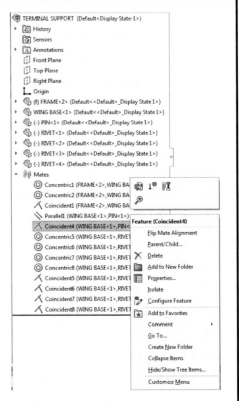

Figure 8-17. Deleting the Coincident Mate between the Pin and the Wing Base.

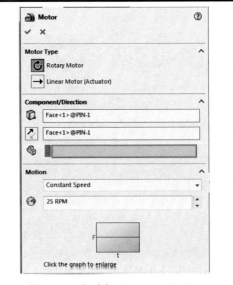

Figure 8-18. Rotary Motor Options.

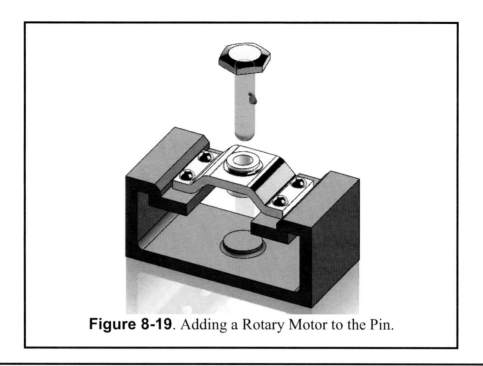

Figure 8-19. Adding a Rotary Motor to the Pin.

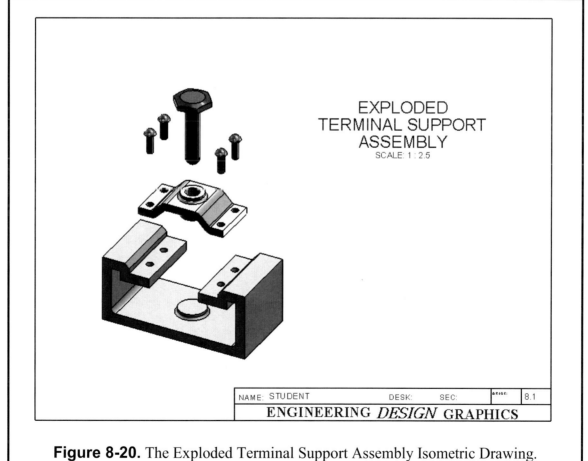

EXPLODED
TERMINAL SUPPORT
ASSEMBLY
SCALE: 1 : 2.5

NAME: STUDENT   DESK: SEC: GRADE: 8.1

ENGINEERING *DESIGN* **GRAPHICS**

Figure 8-20. The Exploded Terminal Support Assembly Isometric Drawing.

Exercise 8.2: Exploded Animation of the PULLEY ASSEMBLY

In Computer Graphics Lab 5 you created the Pulley Assembly. This assembly will be used in this Exercise 8.2.

EXPLODING THE ASSEMBLY

Read through the Kinematic Animation portion of Unit 8 (Pages 8-1 through 8-4) before you begin this exercise.

Open the **Swivel Eye Block** file from your designated folder. You will create the animation by first exploding the assembly. To do this, choose the **Insert** menu and click on **Exploded View** (or click the Exploded View command on the Assembly Toolbar). The "Explode" properties are shown in the earlier **Figure 8-2 on page 8-1**.

The first part to explode is the Big Rivet. Select the **Big Rivet** part of the model; it is added to the Explode Component selection list. Now select the **(Z) direction arrow** to get the correct explode direction. Enter the explode **Distance** of **6.00** inches. If needed, toggle the **Direction** button so the Big Rivet goes to the left. To finish, click the **Add Step** button and *Explode Step 1* is complete. The explode settings for the Big Rivet are shown in **Figure 8-21**. You should now see the Big Rivet move 6 inches to the left from the assembly.

In the same manner, you will now explode the pulley downward. Select the Pulley and the **(Y) direction arrow** to get the correct direction. Enter the explode **Distance** of **6.00** inches. Toggle the **Direction** button to reverse explode direction. To finish, click the **Add Step** button and *Explode Step 2* is complete. The explode settings for the Pulley are shown in **Figure 8-22**. You should now see the Pulley move down 6 inches from the assembly.

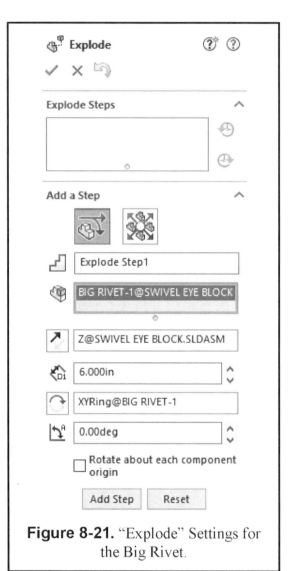

Figure 8-21. "Explode" Settings for the Big Rivet.

You can explode the remaining parts by repeating the above steps, one-by-one, for each part. These steps are as follows.

- Select the component to explode.
- Select the correct direction arrow.
- Enter the explode **Distance** value.
- If needed, toggle the **Reverse Direction** button to get the correct explode direction.
- Click the **Add Step** button to complete the explode step.

Use the following data table for completing this Pulley Assembly explosion process. Note that the Spacer will stay in place and will not move in this explosion process.

Step	Component	Direction	Distance Value
Explode Step 3	Small Rivet 1	Leftward	6.00 inches
Explode Step 4	Small Rivet 2	Leftward	6.00 inches
Explode Step 5	Front Base Plate	Leftward	2.00 inches
Explode Step 6	Back base Plate	Rightward	3.00 inches
Explode Step 7	Eye Hook	Upward	3.50 inches

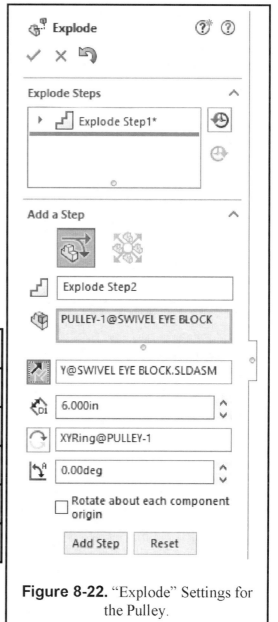

Figure 8-22. "Explode" Settings for the Pulley.

Once this exploded view is completed, click the **OK** button to close the "Explode" command. You should now have an exploded Swivel Eye Hook assembly as shown in **Figure 8-23**.

CREATING THE ANIMATION

Now you will create the animation using the automatic Animator Wizard. Click on **NEW MOTION STUDY** 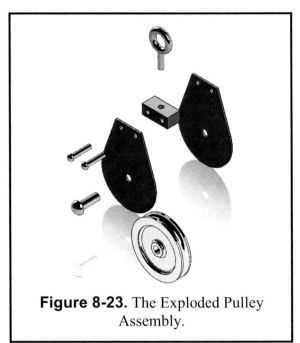 in the Assembly tab of the CommandManager. The "Animator Controller and MotionManager" is displayed on the bottom pane of the screen as shown earlier in **Figure 8-5**, **page 8-2**. Select the **Animation Wizard** command on the Controller (see the choices on **Figure 8-5**). Select the **Explode** option in the "Select Animation Type" dialog, and click **Next** to continue. In the next dialog (see **Figure 8-7**) enter the animation "Duration" of **15** seconds and set "Start Time" to **0**. Finally, click the **Finish** button to create the animation.

Figure 8-23. The Exploded Pulley Assembly.

PLAYING THE ANIMATION

From the "Animation Controller" shown earlier in **Figure 8-8**, press the **Play** button to see the animation. Change to the other animation options. **Play** the animation in the **Normal Mode**. **Play** the animation in the **Loop Mode** and see the assembly move in a repeated fashion until you click **Stop**. **Play** the animation in the **Reciprocate Mode** and see the assembly move in a reciprocating mode until you click **Stop**. Try several combinations of options for this exploded animation. When you have selected the **Normal Mode**, show it to your instructor or lab teaching assistant. **Save** your exploded assembly file as **EXPLODED PULLEY ASSEMBLY.sldasm**.

SAVING THE ANIMATION

Return to the "Animation Controller" and select the **Save Animation** command. The "Save Animation to File" dialog appears on the screen as shown in **Figure 8-24**. Name the animation file **Pulley Assembly** and use an **.AVI file** type. Use the **SOLIDWORKS screen** as the Renderer and set the frames per second rate to **15**. Then click the **Save** button. Click on **OK** in the "Video Compression" dialog box that appears next, in order to accept the default settings. This causes the exploded model animation to be saved to a file as well as played on screen. It may move slowly at this point because it is calculating the motion and saving the video file. Note: You may have to adjust your model viewport to fit the whole animation on your screen as it is being captured.

Figure 8-24. "Save Animation" Menu.

PHYSICAL SIMULATION EXAMPLE

Open the original **Swivel Eye Block** from your designated folder. The assembly components should be completely mated in position. Now you need to remove some of the mating constraints. Expand the **Mates** folder in the FeatureManager Tree, and review all the mates in the **Swivel Eye Block** Assembly. Right click on the **Coincident(X) Spacer<1>, Base Plate<1>** mate, and then select **Delete** as shown in **Figure 8-25**. Answer **Yes** in the **Delete** confirmation dialog. Continue this process and **Delete** <u>all</u> the mates associated with the Front Base Plate, and <u>all</u> the mates associated with the three Rivets.

Select the **Move Component** icon on the Assembly Toolbar, and move all three Rivets away from the assembly. Then move the front Base Plate away from the assembly. See **Figure 8-26** as a preview of the current screen arrangement. Next select the **Motor** icon 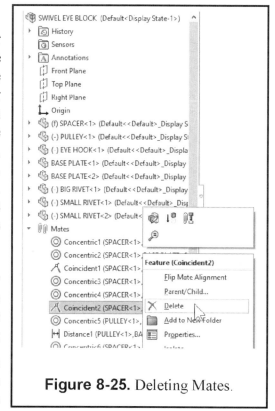 from the Simulation Toolbar (see **Figure 8-9**). Select the hole at the center of the Pulley. Next set the **RPM** below the Constant Speed to 20 RPM as shown in **Figure 8-18** and click OK. Now select the **Calculate Simulation** button on the Simulation Toolbar. The computer will now calculate the physical simulation and play it on the screen as shown in **Figure 8-26**. **If instructed** to do so, <u>show it to your instructor</u> and then press the **Stop or Play** button on the Simulation Toolbar. You can now click the **Play Simulation** button on the Simulation Toolbar if you want.

Now try to add a **Linear Motor** to one of the Rivets or the Eye Hook. You will have to **Delete** a **Coincident** mate holding the Rivet in place to get it to move away from the assembly. Then click the **Calculate Simulation** command to start the new simulation.

Figure 8-25. Deleting Mates.

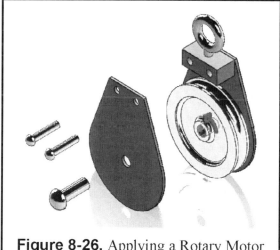

Figure 8-26. Applying a Rotary Motor to the Pulley.

Before you leave the lab, your instructor needs a hard copy of your work. **Open** your previously saved assembly file, **Explode** it, and **Print** an isometric shaded view of the exploded assembly on the Title Block drawing sheet, as shown in **Figure 8-27**. Save your drawing as **Pulley Assembly.slddrw**.

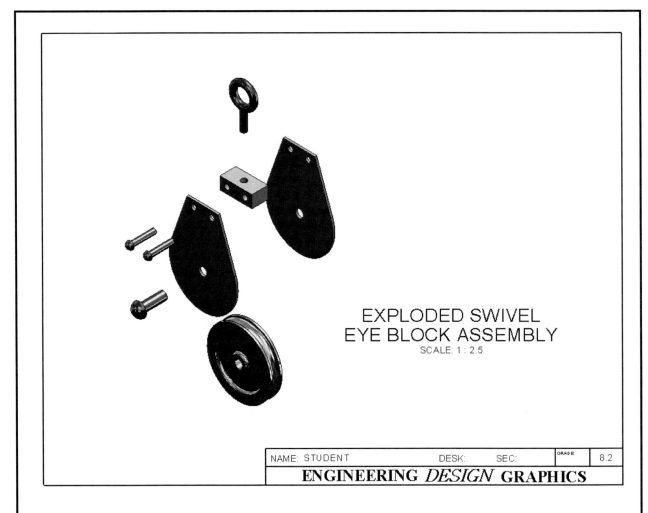

EXPLODED SWIVEL
EYE BLOCK ASSEMBLY
SCALE: 1 : 2.5

| NAME: STUDENT | | DESK: | SEC: | GRADE | 8.2 |

ENGINEERING *DESIGN* GRAPHICS

Figure 8-27. The Exploded Pulley Assembly Isometric View.

Exercise 8.3: Creating Component Drawing Views for Manufacturing

In Exercise 8.3, you will create different component views and add them to a drawing sheet, including special views to show certain details more clearly. Now you will be working in the SOLIDWORKS detail drawing environment.

To begin, open the **Toe Clamp.sldprt** file made in **Exercise 3.4**. Also open your **Title Block Inches.slddrw**. From the **Windows** menu select the **Tile Vertically** command. You will see both open files next to each other. With the drawing file selected, the CommandManager now shows the **View Layout** and **Annotation** tabs with commands needed to make detail drawings.

From the **View Layout** tab select the **Model View** command. Since the Toe Clamp is already open, it is listed in the

Part/Assembly to Insert box. Click on the Next ⮕ button. In the next page, check the **Create Multiple Views** checkbox in the Orientation section, select the **Top** and **Isometric** views, and click **OK** to add the views to the drawing sheet.

Now select the isometric view to make changes to it. In the drawing view's properties, scroll down to the **Display Style** section and select the **Shaded with Edges** option. In the **Scale** section select the **Use Custom Scale**, change the scale to "**User Defined**" and enter **1:3**. Click in the isometric view border or a model edge, and move the view to the upper right corner of the drawing sheet. Also move the top view upward and to the left. Select the Top view, and from the view's properties

Display Style, select the **Hidden Lines Visible** option. From the **Annotations tab** select **Center Mark.** In the Slot

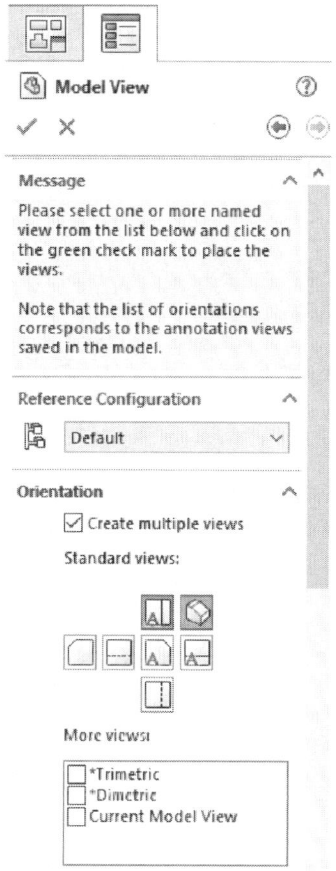

Figure 8-28. Model View Command.

Center Marks section select the "Slot Centers" option, click on the **large arcs** of the slot and click OK. Now select the **Centerlines** command from the Annotations tab. Check the Auto Insert option **Select View** and select the Top view. This will complete the top view as shown in **Figure 8-29**.

Add a Section View

This next step will be covered in greater detail in Unit 9. A section view is created by doing a section line through a model to show internal details. From the **View Layout** tab, select the **Section View** command. In the Section View

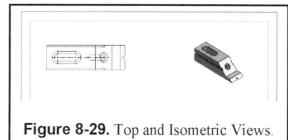

Figure 8-29. Top and Isometric Views.

properties, select the Horizontal Section option, as shown in **Figure 8-30**. Now move your cursor the Top view over either arc of the slot; when it snaps the cutting plane through the center of the slot, click to locate the section line. As you move your cursor below the Top view, the section view will follow your cursor. Click to locate the view in the lower left of the drawing. Now **Right Click** on the **Section View A-A** and from the context menu select **Tangent Edge**, **Tangent Edges Removed**. No hidden lines are to be shown in a full or half section view. If hidden lines are shown in the section view, select the section view and change the Display Style to Hidden Lines Removed in the view's properties. To complete the section view, add the **Centerlines** as we did with the **Top** view.

The cross hatching of section lines is not supposed to be parallel or perpendicular to model lines. Click on one of the section faces to see the Area Hatch/Fill properties, where we can modify it. Uncheck the Material crosshatch box and make the changes in **Figure 8-31**. This will allow you to change the hatch scale and angle.

The section cutting line is the heaviest line used in graphics, so one additional change needs to be made. Go to **Tools, Options, Document Properties**, and under **View Labels** select **Section**. A window will appear that allows you to make changes (**See Figure 8-32**). Under **Line Style** make sure that **Phantom line** is selected, and the line weight is changed to **0.0138**. When you click **OK** you will notice that the cutting line in the top view has changed to a heavy black line.

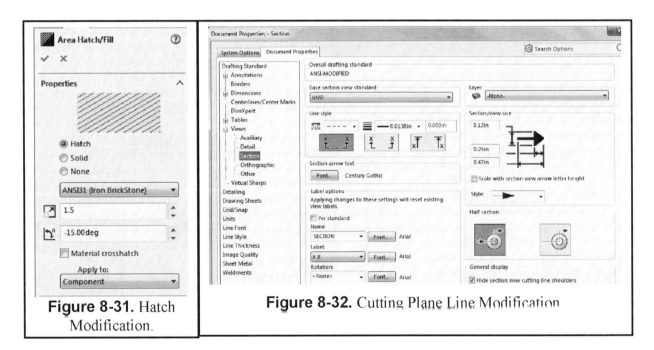

Figure 8-30. Section Cutting Line.

Figure 8-31. Hatch Modification.

Figure 8-32. Cutting Plane Line Modification

Add a Projected View

In order to assure that views are properly aligned and not having to manually align them, they can be added as a projected view, rather than placing them as regular views. You will now place a view to the right of the Section view. From the **View Layout** tab select the **Projected View** command. You will be prompted to select a drawing view from which to project. Select the **Section View** and then move the cursor toward the right of the section view. The right view will appear and follow the cursor. Place the view as shown in **Figure 8-33** and click **OK**. Because you projected from a view with Hidden Lines Removed, the projected view will also have Hidden Lines Removed. Select the right view and change it to Hidden Lines Visible. You should always use Hidden Lines Visible except in full and half sections. As with the other views, add the **Centerline** to complete the view.

Add an Auxiliary View

When you review the views placed on the drawing you will notice that the counterbores in the inclined surface never show as true circles. When dimensioning a drawing it is best to show this feature in a true shape view. This requires you to insert an auxiliary view. From the View Layout tab select **Auxiliary View**. When you do, you will get a prompt to select a Reference Edge to continue. **Select** the **inclined edge** in the section view. This will place an arrow in the direction of sight and allow you to place the auxiliary view in direct alignment with that inclined

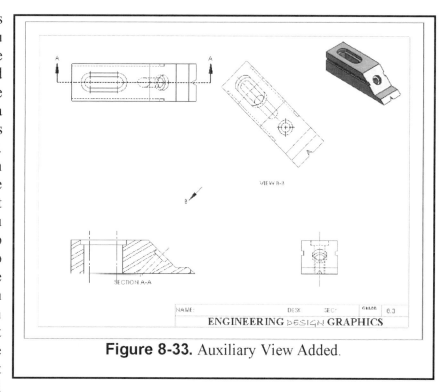

Figure 8-33. Auxiliary View Added.

surface. Change the view to Hidden Lines Visible mode. Also add the **Centerlines**. This now completes the auxiliary and your drawing should be similar to **Figure 8-33**.

Adding a Detail View

Occasionally there are portions of a drawing that are too small to dimension effectively in the standard views; to show these details larger and dimension them clearly, you can create a **Detail View**. From the **View Layout** tab select **Detail View** to show a larger view of the small V-shape of the Toe Clamp in the Top View. Once you have selected the **Detail View** command, go to the top view and draw a small circle around the V. When you move away from the V you will notice an enlarged version of that circle. Place that view in an open area of your drawing. The detail Name and the Scale are displayed below the Detail View. If the detail is too large or too small, you can select the view to change the scale. Your completed drawing should look similar to **Figure 8-34**.

Now save your drawing as **TOE CLAMP PROJECTED VIEWS.slddrw** and print a copy to be turned in to your instructor.

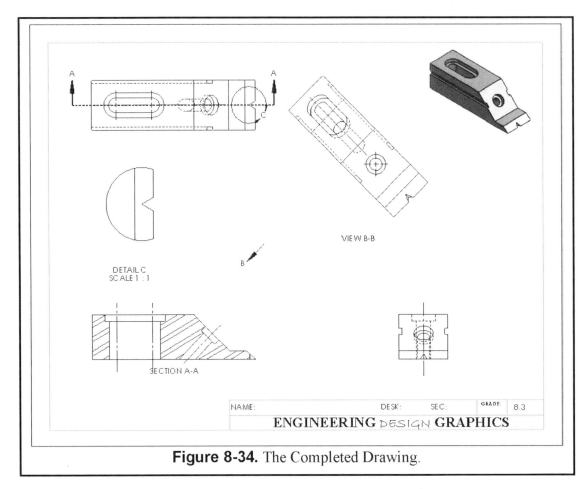

Figure 8-34. The Completed Drawing.

Exercise 8.4: Rapid Prototyping of a Solid Model Part

In this Exercise 8.4, you will build a solid model (or presumably already have one built) as assigned by your instructor. You will then save it in a file format that is the standard in the rapid prototyping industry. This file format is called stereo lithography and is abbreviated .STL. You will then send your .STL file to an available rapid prototyping machine to make a physical prototype of your model.

SAVING THE SOLID MODEL AS A STEREO LITHOGRAPHY (.STL) FILE

You can now build your **New** solid model in SOLIDWORKS, or **Open** a model that has already been built. When you have completed the solid model, do a **Save As**, select "Save as Type" and select **.STL** Files (see **Figure 8-35**). Some example rapid physical prototypes of solid models are shown in **Figure 8-36**.

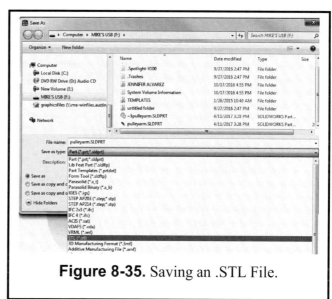

Figure 8-35. Saving an .STL File.

Note: If you need some ideas for solid models for this exercise, examine the accompanying Assignments 8.3.1 to 8.3.4.

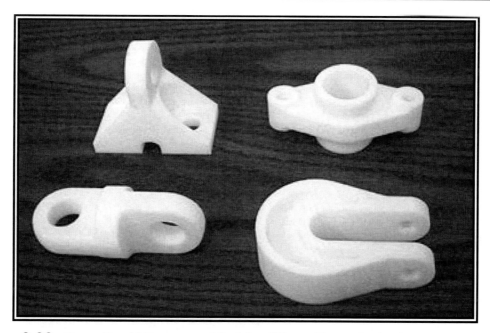

Figure 8-36. Some Rapid Prototype Models of Parts Shown in Assignments 8.3.1 to 8.3.4.

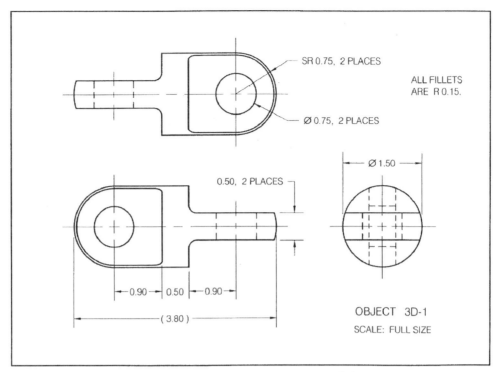

Solid Model Assignment 8.4.1

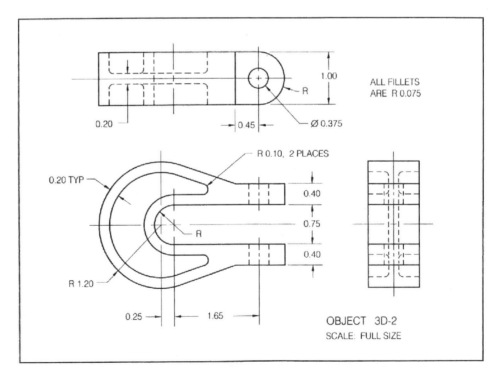

Solid Model Assignment 8.4.2

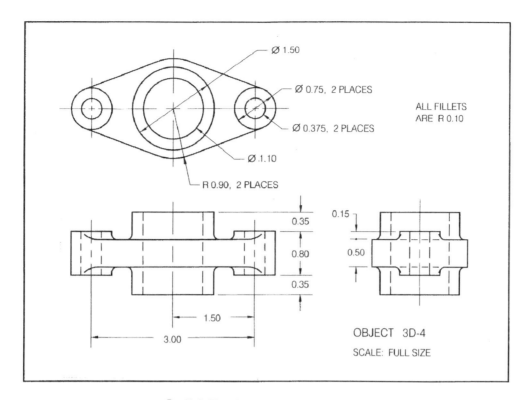

Solid Model Assignment 8.4.3

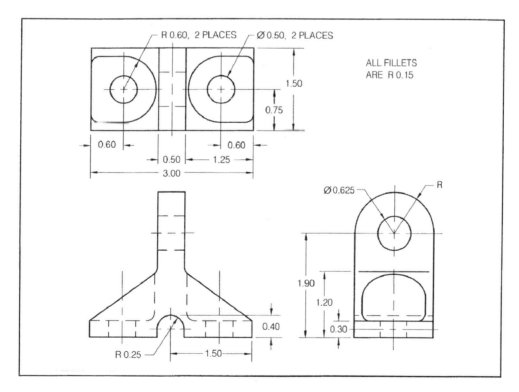

Solid Model Assignment 8.4.4

NOTES:

Computer Graphics Lab 9: Section Views in 3D and 2D

In Computer Graphics Lab 9, you will learn how to make section views from solid models. In each exercise, you will build a 3D solid model of a part. You will learn how to use the "Display Section View" function in SOLIDWORKS to allow you to temporarily get a section of the solid model (e.g. the front half) to visualize the internal features of the 3D part. You will create a new drawing, automatically add orthographic views (front, top, and/or right sides) of the part and use the SOLIDWORKS tools available to create different drawing section views of a part.

USING 3D SECTION VIEWS

The temporary section view of a 3D solid model can be displayed by accessing the **View** menu, then select **Display** and **Section View** as shown in **Figure 9-1**. When you select "Section View", the Section View options are displayed in the PropertyManager as shown in **Figure 9-2**. In the options you can select the cutting plane (e.g. Front plane), the position of the cutting plane relative to the origin, and which side of the model to show. Section views are usually needed to inspect internal features of the model.

Figure 9-1. The Display Section View Command.

You can see the 3D section view of the model in **Figure 9-3**. If you click **OK**, the section view will remain visible. Keep in mind the model has not been cut, it is only a temporary view. To exit the Section View, select the menu **View**, **Display** and deselect the **Section View** command to see the original 3D model.

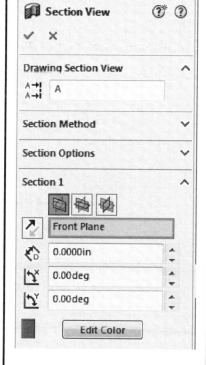

Figure 9-2. The Section View Options.

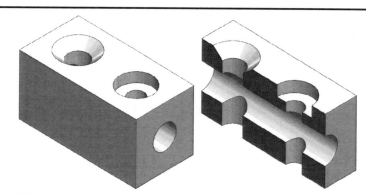

Figure 9-3. A Temporary Section View of a 3D Solid Model.

CREATING 2D SECTION VIEWS ON A DRAWING SHEET

A 2D section view of a model can be created in a drawing. First you must open a **TITLE BLOCK (INCHES or METRIC).drwdot** file. You start by adding model views to a drawing, as we did in **Exercise 8.3**. Once we have added the necessary orthographic model views in the drawing, this then becomes the starting point for making a detailed engineering drawing. From the main views, we can create the necessary section views, and finally add the necessary centerlines, dimensions, notes, and title block information.

First you have to decide which view will be the source for a section view projection. For example, the top view could be the source for projecting a full section in the front view. Select the **Front** view and **Delete** it. From the **View Layout** tab, select the **Section View** command, or the menu **Insert, Drawing View, Section,** as shown in **Figure 9-4**. In the section view properties select

Figure 9-4. Adding a Section View to a Drawing Sheet.

the type of section to create, and click to define the location of the section line in the **Top** view. After the section line is located, a dynamic section view is displayed following the mouse cursor, parallel to the section line. Finally locate the **Section View** in the drawing as shown in **Figure 9-5**.

Depending on which side of the source view you locate the section view, you have the option to reverse the direction using the "Flip Direction" option in the section view properties. SOLIDWORKS will also automatically add the necessary annotations to identify the section line (e.g. A-A) and the section view (e.g. Section A-A).

For example, a Top view of the part shown in **Figure 9-3** was added to a drawing sheet, and from it a horizontal section view was created and placed below it using the above procedure. The resultant drawing for this example is shown in **Figure 9-5**.

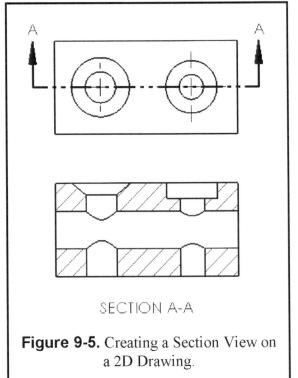

Figure 9-5. Creating a Section View on a 2D Drawing.

Exercise 9.1: ROD BASE SECTION VIEW

In Exercise 9.1, you will build the **Rod Base** part using SOLIDWORKS commands that you have learned in previous labs. You will display a 3D section view of the model to visualize the internal features of the model. Next you will create a 2D drawing with a Top, Isometric and a Section view. To get started, you will first build the solid model part.

BUILDING THE ROD BASE

Open your **ANSI-INCHES.prtdot** in SOLIDWORKS. Immediately **SAVE AS – ROD BASE.sldprt**. Select the **Front** plane and start a new **Sketch**. Draw a vertical **Centerline** through the origin. Sketch the profile shown in **Figure 9-6** using the **Line** tool, and add the given dimension using the **Smart Dimension** tool.

Select the **Revolve Base** command and revolve the profile using a **360** degrees angle. This will be the base feature of the model, as shown in **Figure 9-7**.

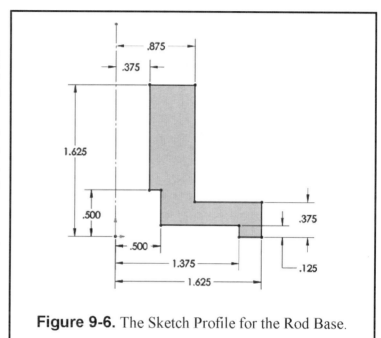

Figure 9-6. The Sketch Profile for the Rod Base.

Now add a **Chamfer** to the *top inner edge* of the through hole. Use the **Angle-Distance** option, set the distance to **0.125** inches and the angle to **45 degrees**.

Next add a **0.0625** inch **Fillets** to the following edges.

1. *Top outer edge* of the upright cylinder.
2. *Intersection edge* between the upright cylinder and the base plate.
3. *Top outer edge* of the base plate.

After you add the fillets, your part should look like **Figure 9-7**.

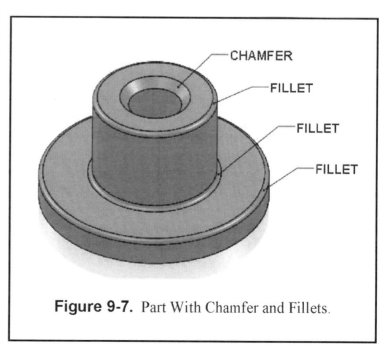

Figure 9-7. Part With Chamfer and Fillets.

Now change to a **Top** view, select the large round face of the model and add a new **Sketch**. Draw a **Circle** and align it with the origin. **Dimension** its diameter to be **0.2500** inches and its distance from the origin to be **1.1250** inches. Now use the **Circular Sketch Pattern** tool to create **4** equally spaced circles (total angle = 360) along this base plate, as shown in **Figure 9-8**. Now make an **Extruded Cut** and use the **Through All** end condition to create four through holes on the base plate.

The final feature to add is the small pinhole that goes through one side of the upright cylinder. Change to a **Right** view orientation, select the

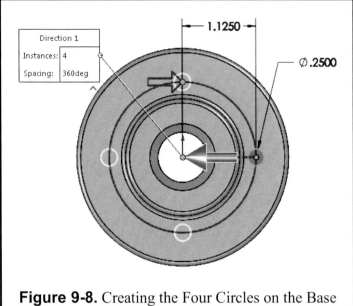

Figure 9-8. Creating the Four Circles on the Base Plate.

Right plane in the FeatureManager and add a new **Sketch**. *Note:* The right plane is in the middle of the hollow cylinder, so you are not actually sketching on any face of the part. Draw a **Circle** on this right plane, aligned vertically with the origin. **Dimension** its diameter to be **0.1875** inches and its distance from the bottom surface of the base to be **1.250** inches, as shown in **Figure 9-9**. Use the **Extruded Cut** command to make the hole through the right side of the cylinder wall using an **Up to Surface** end condition in "Direction 1," where the "up to" surface selected is the outer surface of the upright cylinder. *Note*: You could also just use a **Through-All** end condition for the hole. **Right Click** on **Material (not specified)** in the FeatureManager Tree. Expand **Copper Alloys** and select **Manganese Bronze**. Click on **Apply** and **Close**. Now your model of the Rod Base is complete, and it should look like **Figure 9-10** in an **Isometric** view. **Save** the part in your designated folder and name it **ROD BASE**.

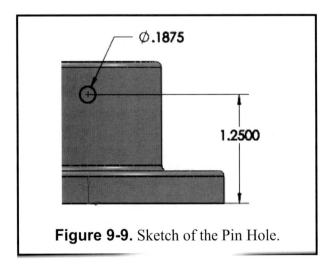

Figure 9-9. Sketch of the Pin Hole.

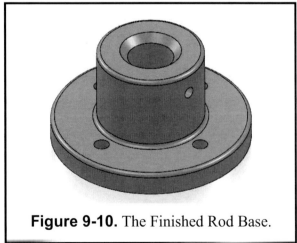

Figure 9-10. The Finished Rod Base.

MAKING A 3D SECTION VIEW OF THE ROD BASE

You will now make a 3D section view of the **Rod Base** model. First select the **Front** plane in the FeatureManager; it will be the section plane. Next, select the menu **View**, **Display**, and then select **Section View**. The "Section View" properties are displayed (refer back to **Figure 9-2**). Leave the "Section Position" at **0.000** inches (middle of model); if needed, click on the **Reverse Section Direction** button, 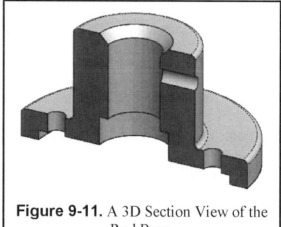 located to the left of the section plane selection box. Click **OK** to finish the section view. Now you have a 3D section view of the Rod Base showing a cut-away view of the part, as shown in **Figure 9-11**.

Figure 9-11. A 3D Section View of the Rod Base.

Now return to the **View** menu, select **Display**, and de-select **Section View** to turn **off** the 3D section view, or select the **Section View** command. _Do not_ **Close** the part yet; it will be needed for the next phase of this exercise, but **SAVE** your progress.

CREATE A DRAWING OF THE ROD BASE

At this point you will make a 2D drawing and add different views of your model. Open the file **TITLEBLOCK-INCHES.drwdot** from your folder and immediately **SAVE AS – ROD BASE.slddrw**.

Before you add the drawing views, there are some general settings that are advisable to make. Select the **Tools**, **Options** menu. In the **Drawings**, **Display Style** section select the "Hidden lines visible" and Tangent Edges "Removed" option as shown in **Figure 9-12**. Next select the **Drawings**, **Area Hatch** tab. Select the **ANSI31** (Iron Brick Stone) "Pattern" option. Set the "Scale" to **1.000** and the Angle" to **0** degrees, as shown in **Figure 9-13**.

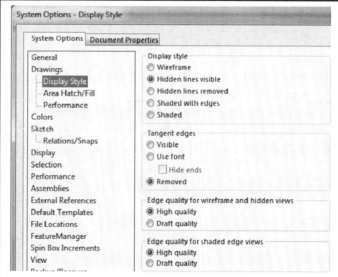

Figure 9-12. Setting Some Drawing Display Options.

In your drawing sheet, select the **View Palette** tab in the **Task Pane** on the right side of the screen. If the **Task Pane** is not visible, right click in any toolbar, and activate the **Task Pane** toolbar.

In the **View Palette** you can see a preview of each orientation of the part open in **SOLIDWORKS**, or you can select a different part from the drop-down list at the top of the **View Palette**. Drag the **Front** view to the lower left of the drawing sheet; immediately after the view is added, move the mouse around to project additional views. Move up to add a  **Top** view; then move up and right, and click to add an **Isometric** view. When done click **OK** to finish projecting views. Since the Front view will be replaced by a section of the **Top** view, select the **Front** view and **delete** it. The views will display using the third angle projection defined in your template, as displayed in **Figure 9-14**. You may have to right click on Sheet 1 in the FeatureManager and select Edit Sheet for the views to show on the drawing sheet. Select the **Isometric View,** and in its properties make sure the **scale** is set to **1:2,** and Display Style is set to **shaded with edges** mode.

Right Click in the isometric view, select **Tangent Edge** and **Tangent Edges Visible**. Move the view to the upper right-hand corner of the drawing sheet. Now select the **Top** view and set its scale to **1:1** and display style to **Hidden Lines Visible**.

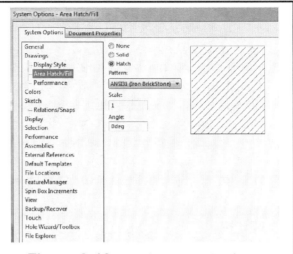

Figure 9-13. Hatch Pattern Settings.

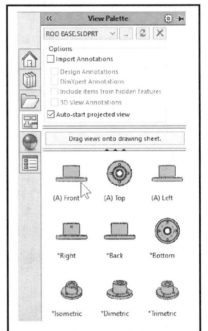

Then move the image to the upper left of the drawing sheet.

ADDING A SECTION VIEW OF THE ROD BASE

The Top view is the source view to create a section view. In your drawing sheet, select the **Section View** command from the **View Layout** tab. In the Section View properties select the **Horizontal** section line option. (See **Figure 9-15**.)

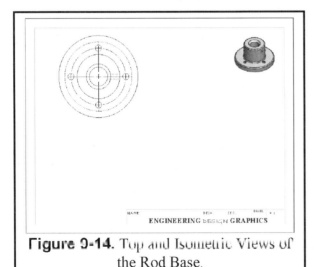

ENGINEERING DESIGN GRAPHICS

Figure 9-14. Top and Isometric Views of the Rod Base.

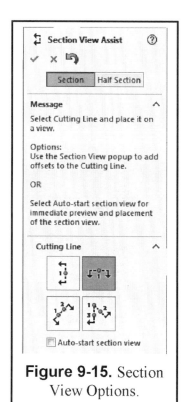

Figure 9-15. Section View Options.

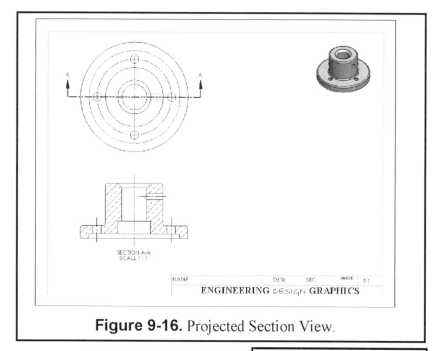

Figure 9-16. Projected Section View.

Figure 9-17.

Move your cursor to the center of the **Top View**, when the **Section Line** snaps to the center, click to locate it, move your cursor down to the place where you want to position your Section view and click to finish the section view command.

Notice the section view is aligned vertically with the top view, and parallel to the section line. The section line is automatically labeled **A-A,** and the section view is labeled **Section A-A**, as indicated in **Figure 9-16**. If the section line arrows are pointing down, click in the **Flip Direction** button. If Hidden lines are showing in the Section View, change it to Hidden Lines Removed. Finally, select the menu **Insert – Annotations – Centerline** and check the box in front of the **Select View** to add the centerlines to the Section View.

If desired, increase the font size of the "Section A-A" label to **18 point**.

Now you need to add a Right view. From the **View Layout** tab, select the **Projected view** command. Select the section view and move to the right to project the **Right** view. When you reach the desired location, click to position the view. By projecting the view, the **Right** view will be aligned horizontally with the section view.

Change the **Right** view to **Hidden Lines Visible**. See **Figure 9-18** for an example of the finished drawing. From the **Annotation** tab select **Note** to add a **ROD BASE** title and **SCALE 1:1** to the drawing as shown in **Figure 9-18**. Use **all capitals**, **Arial** font for all the labels. Use **24 point** font for the "**ROD BASE**" title and **12 point** font for all the other annotations.

When you are finished **Save** your drawing as **ROD BASE.slddrw** in your designated folder (*note*: the drawing's extension is **.slddrw**). The finished drawing, with title block and isometric view, is shown in **Figure 9-18**. **Print** a hard copy to submit to your instructor.

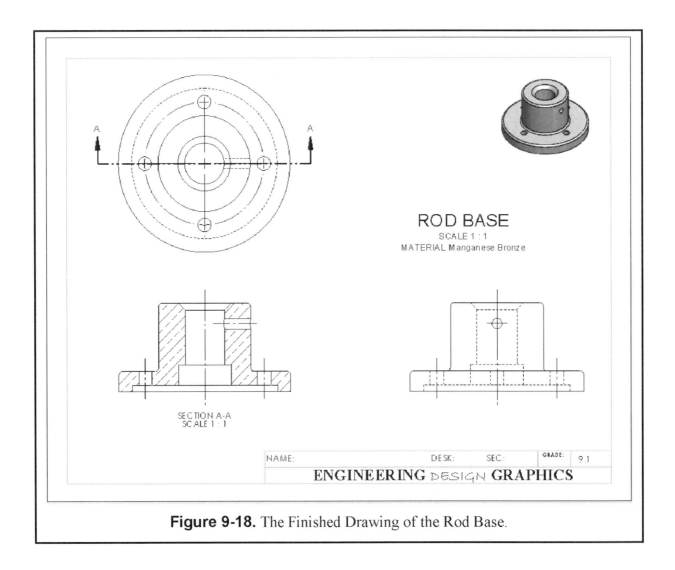

Figure 9-18. The Finished Drawing of the Rod Base.

Exercise 9.2: TENSION-CABLE BRACKET SECTION VIEW

In this Exercise 9.2, you will build a solid model of the Tension Cable Bracket using SOLIDWORKS commands that you have learned in previous labs. You will make a 3D section view of the model to visualize the internal features of the model. Next you will create a 2D drawing including a section view of the part. To get started, you will first build the solid model part.

BUILDING THE TENSION CABLE BRACKET

Open your **ANSI-INCHES.prtdot** in SOLIDWORKS. Immediately **SAVE AS – TENSION-CABLE BRACKET.sldprt**. Go to **Tools – Options – Document Properties – Units** and set your decimal places to **Three**. Select the **Top** plane and start a new **Sketch**. Draw a **Center Rectangle** centered at the origin as indicated in **Figure 9-19**. Add the given dimension using the **Smart Dimension** tool. Select the **Extrude Base** command and extrude it **0.250** inches upward from the plane. Rename this feature as "Base plate."

Now select the top face of the base plate, add a new **Sketch** and draw a **Circle** on the top plate and **Dimension** it as shown in **Figure 9-20**. Add a horizontal relationship between the hole and the origin. Draw a vertical **Centerline** and **Mirror** the circle to the left side. Next, make an **Extruded Cut** using the **Through All** end condition with a **Draft** angle of **15** degrees, as shown in **Figure 9-21**. This results in the two tapered holes in the base plate.

Now start another **Sketch** on the top face of the Base Plate. Draw another **Circle**, centered at the *origin* and with a *diameter* of **1.750** inches, make an **Extruded Boss** with the following information:

"Direction 1" = **Blind 0.875** inches
"Direction 2" = **Blind 0.875** inches

Now you now have a boss going in each direction.

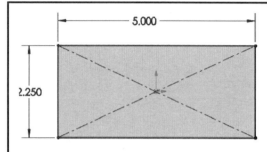

Figure 9-19. The Initial Sketch Profile.

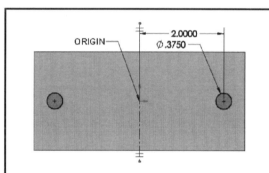

Figure 9-20. Sketching the Two Plate Holes.

Figure 9-21. Tapered Hole Settings.

For the next feature change to a Top view, add a **Sketch** on the top face of the boss, and draw a **Circle** at the origin and dimension it **1.000** inch in diameter. Add an **Extruded Cut** using the **Through All** end condition with a **Draft** angle of **7.5** degrees. This makes a tapered hole through the part seen with Hidden Lines Visible in **Figure 9-22**.

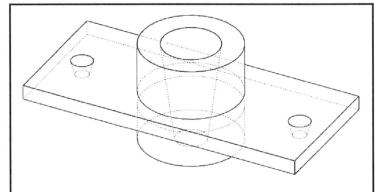

Figure 9-22. Cable Bracket with Hidden Lines Visible at This Stage in the Design Process.

You will now add triangular supports to the boss feature. These will be offset from the center, so you first need to add a reference plane. First select the **Front** plane in the FeatureManager. From the **Features** tab select **Reference Geometry**, **Plane**. Make the new Plane **0.750** inches parallel to and in front of the Front plane, and click OK to finish. The new Plane1 is now added to the FeatureManager. Select **Plane1** and add a new **Sketch**. Change to a Front view, draw a triangle and **Dimension** it as shown in **Figure 9-23**. Draw a vertical **Centerline** in the origin, and **Mirror** the triangle about it.

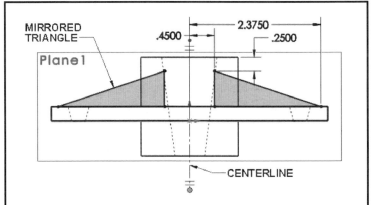

Figure 9-23. Sketching and Mirroring a Triangular Support on Plane1.

Now make an **Extruded-Boss** with this sketch, using a **Blind** distance of **0.250** inches towards the center of the part (the origin). You now have two triangular supports on the front side of the object. From the Features tab select the Mirror command, and mirror the triangle extrusions about the **Front** plane. Your part now looks like **Figure 9-24**.

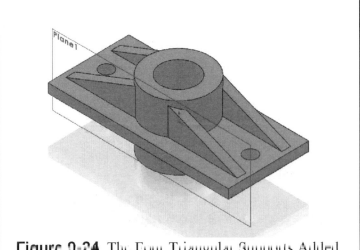

Figure 9-24. The Four Triangular Supports Added.

To complete the Tension Cable Bracket, add the following features. First add a **Chamfer** to the inner edge of the hole on top of the upright boss. Make the chamfer using a **Distance-Distance** option with both distances equal to **0.125** inches. Next we need to add **Fillets** to multiple edges. Make them all with a radius of **0.0825** inches. Select the following edges to fillet with this command:

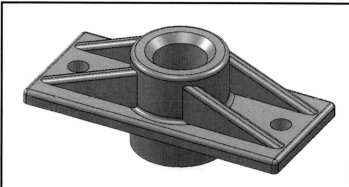

Figure 9-25. The Finished Tension Cable Bracket after Adding the Chamfer and Fillets.

- Outer top edge of upright boss.
- All intersections of the upright boss with the base plate, both above and below the base plate.
- The top outer edges of the base plate.
- The top edges of the four triangular supports.
- The vertical edges of the rectangular base.

When you are finished, your part will look like **Figure 9-25** in a **Trimetric** view. Be sure to **Save** your part with the name **TENSION CABLE BRACKET.sldprt**.

MAKE A 3D SECTION VIEW OF THE TENSION CABLE BRACKET

You will now make a 3D section view of the Cable Bracket model. First select the section plane. Select the **Front** plane in the FeatureManager, and select the menu **View**, **Display**, **Section View**. In the "Section View" properties (refer back to **Figure 9-2**) leave the "Offset Distance" at **0.000** inches, and if needed, click in the **Reverse Direction** button to flip the section's view to match **Figure 9-26**. Click **OK** to finish. You should now have a 3D section view of the Cable Bracket showing a cut-away view of the part using the **Front** plane, as shown in **Figure 9-26**.

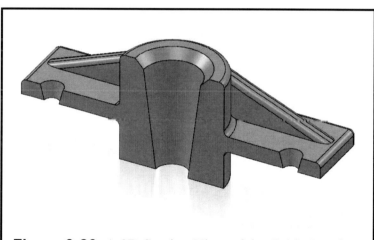

Figure 9-26. A 3D Section View of the Cable Bracket.

To turn **off** the Section View display mode, select the menu **View, Display, Section View** and de-select it. _Do not_ **Close** your solid model file yet because you will need it for the next exercise.

CREATE A DRAWING OF THE CABLE TENSION BRACKET

At this point you will make a 2D drawing and add different views of your model. Open the file **TITLEBLOCK-INCHES.drwdot** from your folder and immediately **SAVE AS – CABLE TENSION BRACKET.slddrw**.

Before you start the drawing views, there are some general settings that are advisable to make. Select the **Tools**, **Options** menu. In the **Drawings**, **Display Style** section select the "Hidden lines visible" and Tangent Edges "Removed" option as shown in **Figure 9-27**. Next select the **Drawings**, **Area Hatch** tab. Select the **ANSI31** (Iron Brick Stone) "Pattern" option. Set the "Scale" to **1.000** and the Angle" to **0** degrees.

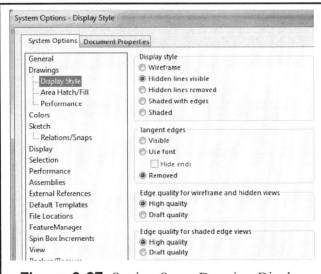

Figure 9-27. Setting Some Drawing Display Options.

In your drawing sheet, select the **View Palette** tab in the **Task Pane** on the right side of the screen.

Open the **View Palette** where you can see a preview of each orientation of the part open in **SOLIDWORKS**, or you can select a different part from the drop-down list at the top of the **View Palette**.

In the View Palette turn off the **Auto-Start Projected View** option at the top, and drag the **Top** view to the upper left of the drawing sheet, and the **Isometric** view to the upper right side of the drawing sheet.

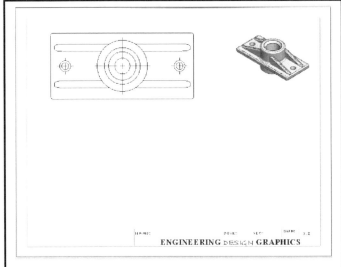

Figure 9-28. The Top and Isometric Views of the Cable Bracket.

Select the **Isometric View** and from the view's properties change the **Scale** to 1:2 and the Display Style to **Shaded with Edges** mode.

Now select the **Top view,** change its **Scale to 1:1** and select the Hidden Lines Visible display mode. Move the views to match the image on **Figure 9-28**. If needed, right click on the Top view and from the context menu select the **Tangent Edges, Tangent Edges Removed** option.

ADD A SECTION VIEW OF THE CABLE BRACKET

To add a section view of the **CABLE BRACKET**, select the **Section View** command from the **View Layout** tab. In the section view's properties select the **Horizontal** section **(See Figure 9-29)**. Move your cursor to the center of the Top view; when the Section Line snaps to the center, click to locate it, move your cursor down to the place where you want to position your section view and click to finish the section view command.

Remember the section view is aligned vertically with the top view, and parallel to the section line. The section line is automatically labeled **A-A,** and the section view is labeled **Section A-A**, as indicated in **Figure 9-30**. If the section line arrows are pointing down, click in the **Flip Direction** button. If Hidden Lines are showing in the Section View, change it to Hidden Lines Removed. Finally, select the menu **Insert – Annotations – Centerline** and check the box in front of the **Select View** to add the centerlines to the Section View.

If desired, increase the font size of the "Section A-A" label to **18 point**.

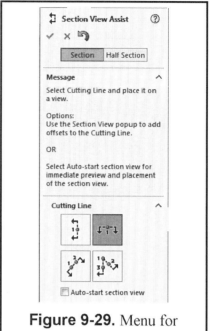

Figure 9-29. Menu for Section Lines.

Now you need to add a Right view. From the **View Layout** tab, select the **Projected view** command. Select the section view and move to the right to project the **Right** view. When you reach the desired location, click to position the view. By projecting the view, the **Right** view will be aligned horizontally with the section view.

Change the **Right** view to **Hidden Lines Visible**. See **Figure 9-18** for an example of the finished drawing. From the **Annotation** tab select **Note** to add a **ROD BASE** title and **SCALE 1:1** to the drawing as shown in **Figure 9-18**. Use **all capitals, Arial** font for all the labels. Use **24 point** font for the "**ROD BASE**" title and **12 point** font for all the other annotations.

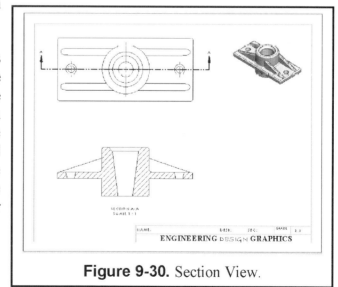

Figure 9-30. Section View.

When you are finished **Save** your drawing file as **TENSION CABLE BRACKET.slddrw** in your designated folder. The finished drawing, with title block and isometric view, is shown in **Figure 9-31**. **Print** a hard copy of the drawing to submit to your instructor.

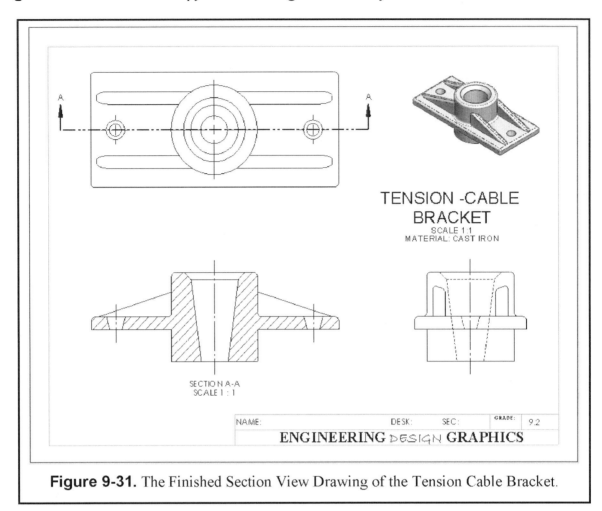

Figure 9-31. The Finished Section View Drawing of the Tension Cable Bracket.

Exercise 9.3: MILLING END ADAPTER
Section Views

In Exercise 9.3, you will model a Milling End Adapter using SOLIDWORKS commands that you have learned in previous labs. You will display a 3D section view of the model to visualize the internal features of the model. Next you will create a 2D drawing with a 2D section view. To get started, you will first build the solid model part.

BUILDING THE MILLING END ADAPTER

Open your **ANSI-INCHES.prtdot** in SOLIDWORKS. Immediately **SAVE AS – MILLING END ADAPTER.sldprt**. Select the **Front** plane and add a new **Sketch**. Draw a horizontal **Centerline** through the origin. Sketch and dimension the profile shown in **Figure 9-32**. If needed, add a horizontal relation between the top endpoints of the 0.125" wide slots.

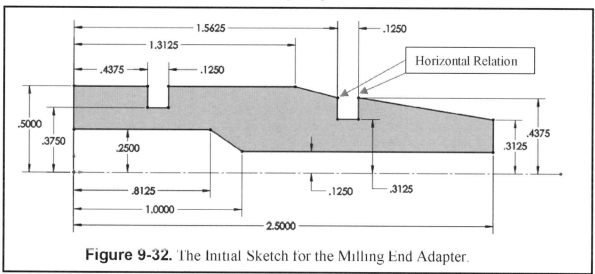

Figure 9-32. The Initial Sketch for the Milling End Adapter.

Select the **Revolve Base** command and revolve the profile into a solid using a **360** degrees angle. Now add a **Chamfer** to the end edges of the part. Use the **Distance-Distance** option, with both distance values being **0.0625** inches. At this stage, you should have a revolved solid part with the ends chamfered, as shown in **Figure 9-33**.

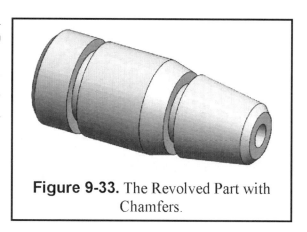

Figure 9-33. The Revolved Part with Chamfers.

You now need to create two screw holes on the round outer surfaces of the Milling End Adapter. SOLIDWORKS needs a flat plane to sketch on, so first select the **Top** plane in the FeatureManager. From the **Features** tab select the **Reference Geometry**, **Plane** command. In the **Plane** parameters select the Offset Distance option and make the new plane **0.5000** inches *above* the Top plane as shown in **Figure 9-34**. Click **OK** to finish the reference plane.

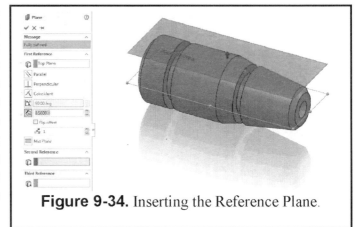

Figure 9-34. Inserting the Reference Plane.

Select **Plane1** and add a new **Sketch** in it. Change to a **Top** view orientation to better see the sketch plane, and draw two **Circles** and **Dimension** them as shown in **Figure 9-35**. Add an equal geometric relation to make both circles the same, and dimension the diameter of either one **0.1250** inches. The first circle center is **0.3125** inches from the **Origin** on the left side of the Milling Adapter, and the second circle center is **0.3750** inches to the right of the first circle center. To fully define the sketch, add a horizontal relation between the circles' center and the Origin.

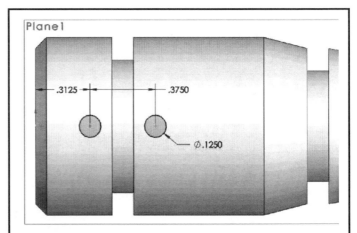

Figure 9-35. Adding Two Circles for the Screw Holes.

Now make an **Extruded Cut** down through the cylinder wall using an **Up to Next** end condition in "Direction 1." The **Up to Next** end condition will make the cut until it finds the next model face, in this case the *inner* face of the Milling Adapter. This results in the holes only going through the top half of the part. Now your model of the Milling End Adapter is complete, and it should look like **Figure 9-36**. To complete the model, assign the **Manganese Bronze** material from the **Copper Alloys** library. **Save** the part in your designated folder as **MILLING END ADAPTER.sldprt**.

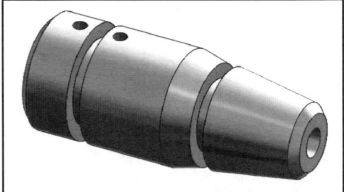

Figure 9-36. Finished Milling End Adapter Part.

MAKING A 3D SECTION VIEW OF THE MILLING END ADAPTER

You will now make a 3D section view of the Milling End Adapter model. First select the **Front** plane in the FeatureManager; it will be the section plane. Next, select the menu **View**, **Display**, and then select **Section View**. The "Section View" properties are displayed (refer back to **Figure 9-2**). Leave the "Section Position" at **0.000** inches (middle of model); if needed, click on the **Reverse Section Direction** button, 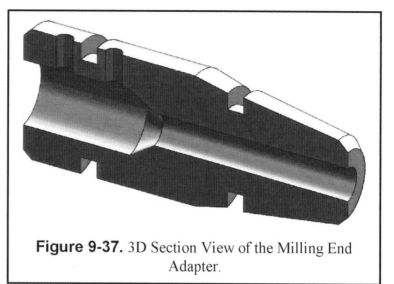 located to the left of the section plane selection box. Click **OK** to

Figure 9-37. 3D Section View of the Milling End Adapter.

finish the section view. Now you have a 3D section view of the Milling End Adapter showing a cut-away view of the part, as shown in **Figure 9-37**.

Now return to the **View** menu, select **Display**, and then de-select **Section View** to turn **off** the 3D section view, or select the **Section View** command. *Do not* **Close** the part, it will be needed for the next phase of this exercise, but **SAVE** your progress.

CREATE A DRAWING OF THE MILLING END ADAPTER

At this point you will make a 2D drawing with three different views of your model. Open the file **TITLEBLOCK-INCHES.drwdot** from your folder and immediately **SAVE AS – MILLING END ADAPTER.slddrw**.

Before you add the drawing views, there are some general settings that are advisable to make. Select the **Tools**, **Options** menu. In the **Drawings**, **Display Style** section select the "Hidden lines visible" and Tangent Edges "Removed" option as shown in **Figure 9-38**.

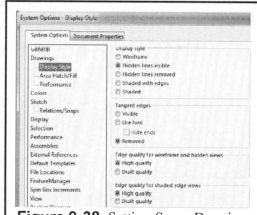

Figure 9-38. Setting Some Drawing Display Options.

Note: The options listed in the **System Options** tab control the behavior of the entire SOLIDWORKS software, regardless of the model or drawing we are working on, including Performance, Display, Selection, File Locations, etc.
The options listed in the **Document Properties** only affect the document we are working on, and they include the Drafting Standard (which controls annotation display), Units, Material, etc.

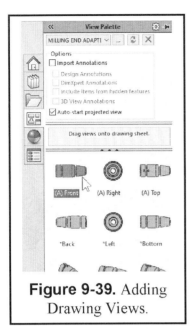

Figure 9-39. Adding Drawing Views.

In your drawing sheet, select the **View Palette** tab in the **Task Pane** on the right side of the screen.

Open the **View Palette** where you can see a preview of each orientation of the part open in **SOLIDWORKS**, or you can select a different part from the drop-down list at the top of the **View Palette**. Also, make sure the **Auto-Start projected view** option is selected.

Now drag the **Front** view to the lower left of the drawing sheet; immediately after the view is added, move the mouse around to project the additional views. Move up to add a **Top** view; then move up and right, and click to add an **Isometric** view, move to the right and click to add the **Right** view. The new views will be displayed using the third angle projection defined in your template, as displayed in **Figure 9-40**.

After the views are added, you may see that the drawing views are too small. To change the drawing sheet's scale, go to the FeatureManager, right click on **Sheet1** and select **Properties** from the context menu. Set the Scale to **2:1** and click **OK** as shown in **Figure 9-40**. Optionally, you can select the views individually and set a custom scale in the view's properties. Select the **Isometric** view, and change it to **Shaded With Edges** and a 1:1 scale. Add the missing **Centerlines** to the three main views.

ADD A BROKEN-OUT SECTION VIEW OF THE END ADAPTER

Broken-out Section

A **Broken-Out section** shows a partial section to a defined depth in a view. To create it, select the **Broken-Out Section** command from the View Layout tab. The **Spline** sketch tool will be automatically activated. Go to the Front view and draw a closed profile to define the area for the section, as suggested in **Figure 9-41**. When the spline (or any other closed profile) is completed, the **Broken-Out Section** properties are shown.

Turn on the **Preview** checkbox to preview the results. In the **Depth** distance box, enter **0.500** inches. This places the cut depth plane through the middle of the part. You will see the depth planes in the **Top** and **Right** views, including direction of cut arrows, as shown in **Figure 9-42**. Click **OK** to finish.

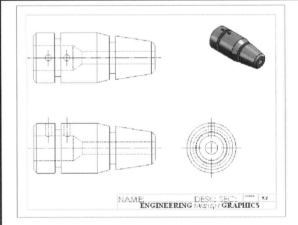

Figure 9-40. Three Views of the End Adapter, Using a 2:1 Scale.

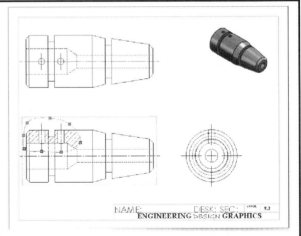

Figure 9-41. Closed Spline on the Front View for the Broken-Out Section.

Note: You can draw the closed spline **beforehand**, and drag the nodes to adjust the section area. When the closed profile covers the area of interest, pre-select the closed profile and then select the **Broken-Out Section** view command.

You now have a **Broken-Out Section** view of the Milling End Adapter in the **Front** view, and **Top**, **Right** and **Isometric** views, as shown in the **Figure 9-43**. If missing, add the necessary **Center Marks** or **Centerlines** from the Annotations tab. Use the **NOTE** command to add a label below the **Front** view that says **BROKEN-**

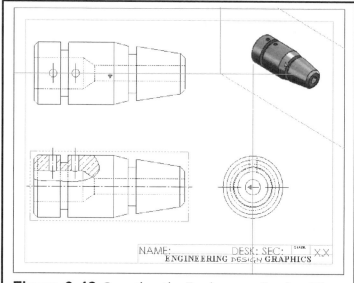

Figure 9-42. Inserting the Broken-out Section View and Previewing It.

OUT SECTION using **14**-point font size. Add a new **Note** that reads: **MILLING END ADAPTER, SCALE 2:1** and **MATERIAL: MANGANESE BRONZE** to the drawing as shown in **Figure 9-44**. Use *caps*, **Arial** font for all the labels, **20-point** font for the "Milling End Adapter" title, and **12-point** font for the other annotations.

When you are finished, **Save** your drawing as **MILLING END ADAPTER.slddrw** in your designated folder. A finished drawing is shown in **Figure 9-43**. **Print** a hard copy to submit to your instructor.

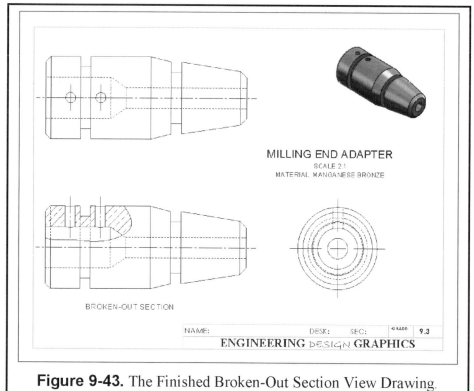

Figure 9-43. The Finished Broken-Out Section View Drawing.

Exercise 9.4: PLASTIC REVOLVING BALL ASSEMBLY Section View

In Exercise 9.4, you will build the three parts of the Plastic Revolving Ball assembly using SOLIDWORKS commands that you have learned in previous labs. You will display a 3D section view of the model to visualize the internal features of the model. Next you will make an assembly drawing and create a 2D section view of the assembly. To get started, build the solid model of each part and then create an assembly of the three parts.

Building the Ball

Open **ANSI-INCHES.prtdot** from your folder. Immediately **SAVE AS – PLASTIC BALL.sldprt.** From the materials library assign **PF** found in the **Plastics** library. Select the **Front** plane and start a new sketch. Draw a horizontal **Centerline** through the **Origin**, and draw the profile shown in **Figure 9-44** with the dimensions provided. Add a geometric relation to make the center of the small arc **Vertical** with the Origin.

Select the **Revolved Boss** command and revolve the sketch **360**-degrees. This creates the Plastic Ball. Your part should look like **Figure 9-45**.

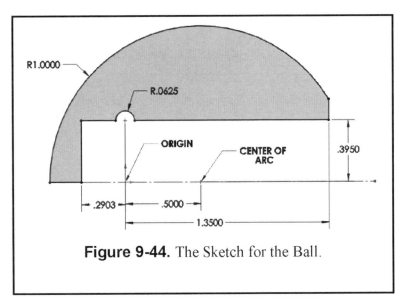

Figure 9-44. The Sketch for the Ball.

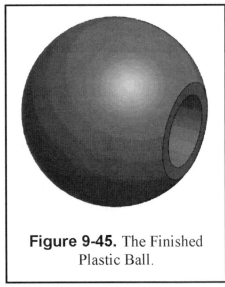

Figure 9-45. The Finished Plastic Ball.

Building the Steel Shaft

Open **ANSI-INCHES.prtdot** from your folder. Immediately **SAVE AS – STEEL SHAFT.sldprt**. From the materials library assign **AISI 316 Annealed Stainless Steel (SS)** found in the **Steel** library. Select the **Front** plane and start a new sketch. Draw a horizontal **Centerline** through the **Origin**, draw and dimension the sketch profile shown in **Figure 9-46**. Select the two horizontal lines at the top, and add a **Collinear** relation between them; this way you only need one dimension.

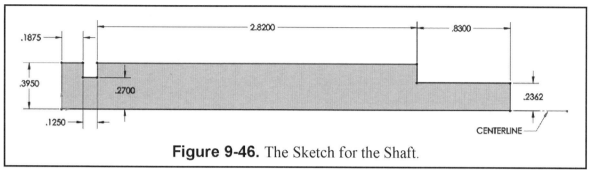

Figure 9-46. The Sketch for the Shaft.

Select the **Revolved Boss** command and revolve the sketch **360**-degrees. This creates the Steel Shaft. Change to a Right view, select the small end of the shaft and add a new sketch. Draw a **Hexagon** as shown in **Figure 9-47** and make a **Cut Extrude** using the **Blind** end condition **0.375** inches deep. Finally add a **0.0375**-inch **Chamfer** on both ends of the shaft. Your part should look like **Figure 9-48**. **Save** your second part as **STEEL SHAFT.sldprt**.

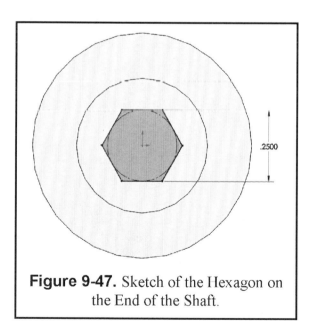

Figure 9-47. Sketch of the Hexagon on the End of the Shaft.

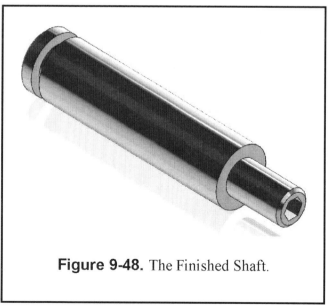

Figure 9-48. The Finished Shaft.

Building the Snap Ring

Open **ANSI-INCHES.prtdot** from your folder. Immediately **SAVE AS – SNAP RING.sldprt**. From the materials library assign **Alloy Steel** found in the **Steel** library. Select the **Front** plane and add a new sketch. Draw and dimension the profile shown in **Figure 9-49** including a horizontal **Centerline** through the **Origin**. Select the origin and the arc's center, and add a **Vertical** geometric relation to **center it above the origin**.

Select the **Revolved Boss** command and revolve the sketch **360**-degrees. This creates the Snap Ring.

Change to a Right view and select the **Right** plane to start a new sketch. Draw the sketch profile shown in **Figure 9-50** and add a **Cut Extrude** using the **Through all - Both** end condition.

Your part should look like **Figure 9-51**. **Save** your part as **SNAP RING.sldprt**.

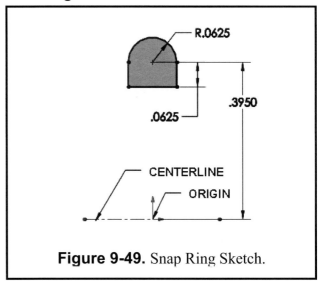

Figure 9-49. Snap Ring Sketch.

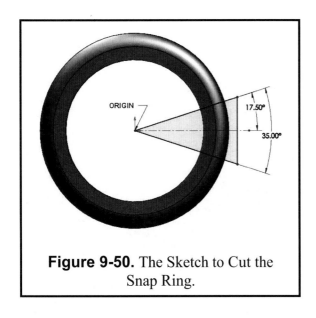

Figure 9-50. The Sketch to Cut the Snap Ring.

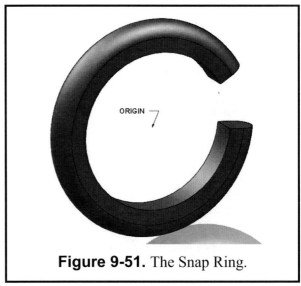

Figure 9-51. The Snap Ring.

Building the Assembly

With the three part files open, go to the menu **File, New, Assembly**. In the new assembly, the **Begin Assembly** command is open. Select the **Snap Ring** from the **Part/Assembly to Insert** list and click **OK**. The **Snap Ring** will be added to the assembly and located in the assembly's origin. From the **Assembly** tab select the **Insert Components** command, select the **Steel Shaft** from the list and locate it in the assembly window anywhere other than the **origin**. Add a **Mate** to make the **Snap Ring** and the **Steel Shaft Concentric**, and then mate one of the flat faces of the Snap Ring **Coincident** to the matching face in the shaft's ring groove. This will position the snap ring in the groove of the steel shaft. When this is completed, repeat the **Insert Components** command, select the **Plastic Ball** and click **OK**; it will be added and aligned with the assembly's origin. **Save** the assembly to your folder as **PLASTIC REVOLVING BALL.sldasm**.

MAKING A 3D SECTION VIEW OF THE PLASTIC REVOLVING BALL ASSEMBLY

You will now make a 3D section view of the Plastic Revolving Ball Assembly. First select the **Front** plane in the FeatureManager; it will be the section plane. Next, select the menu **View, Display**, and then select **Section View**. The "Section View" properties are displayed (refer back to **Figure 9-2**). Leave the "Section Position" at **0.000** inches (middle of model); if needed, click on the **Reverse Section**

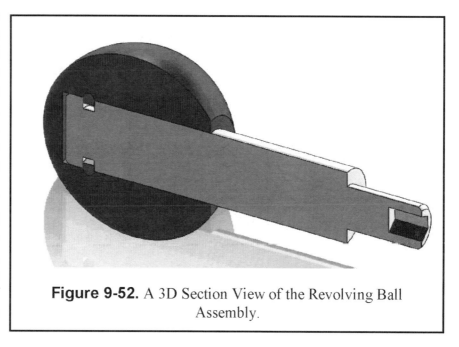

Figure 9-52. A 3D Section View of the Revolving Ball Assembly.

Direction button, located to the left of the section plane selection box. Click **OK** to finish the section view. Now you have a 3D section view of the the **Plastic Revolving Ball** Assembly, showing a cut-away view of the part, as shown in **Figure 9-52**.

Now return to the **View** menu, select **Display**, and de-select **Section View** to turn **off** the 3D section view, or select the **Section View** command. _Do not_ **Close** your assembly yet; it will be needed for the next phase of this exercise. You may close the individual parts at this time.

CREATING A DRAWING OF THE REVOLVING BALL ASSEMBLY

Before you add the drawing views, there are some general settings that are advisable to make. Select the **Tools**, **Options** menu. In the **Drawings**, **Display Style** section select the "Hidden lines visible" and Tangent Edges "Removed" option as shown in **Figure 9-53**.

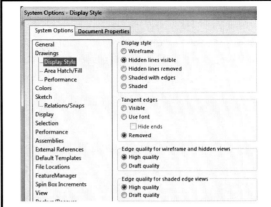

Figure 9-53. Setting Some Drawing Display Options.

Now open the **TITLEBLOCK-INCHES.drwdot** from your folder and **SAVE AS – REVOLVING BALL ASSEMBLY.** In your drawing sheet, select the **View Palette** tab in the **Task Pane** on the right side of the screen. Turn off the Auto-Start projected view option and add the **Top** and **Isometric** views to the drawing. Select the **Isometric View** and if needed, set the view's **Scale** to **1:2** and change the display mode to **Shaded with Edges**. Move the view into the upper right-hand corner of the drawing. Now select the **Top view** and set its **Scale to 1:1** with **Hidden Lines Visible** mode. Move the **Top** view to the upper left of the drawing.

To add a section view of the **REVOLVING BALL ASSEMBLY**, select the **Section View** command from the **View Layout** tab. In the section view's properties select the **Horizontal** section **(See Figure 9-54)**. Move your cursor to the center of the **Top** view, when the Section Line snaps to the center of the view, click to locate it. You will be presented with the Section Scope dialog, asking you to select the components that should not be cut. Click OK to close it and continue to position the section view. Click to locate the view directly below the Top view.

The section line will be labeled **A-A** in the Top view. The section line arrows should point in the correct direction as you move the section view into position. If the arrows are pointed downward, select the flip direction box. The section view is automatically labeled **Section A-A**, as shown in **Figure 9-54**.

Once the section view is placed, you will notice that the section lines may not be at different angles and the scale of the section lines is not appropriate. If you **Right Click** on the section lines, a window appears that lets you edit the **Crosshatch Properties**. Make changes in the Hatch Pattern Scale and the Hatch Pattern Angle to change the density of the Hatch and their directions.

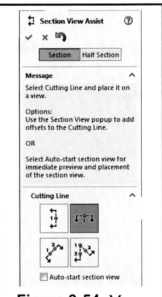

Figure 9-54. Menu for Section Lines.

Now select the **Section View** and click in the **Projected View** command from the **View Layout** tab, and locate the **Right** view under the **Isometric** view. You will notice that centerlines and hidden lines are missing in the **Right** view. Change the Right view to **Hidden Lines Visible** mode, then select **Insert, Annotations, Center Mark** to add the missing **Center Marks** in the **Right** view.

To add the center mark, select a circular edge in the right view and the center mark will be added. Use **24 point** font for the "**PLASTIC REVOLVING BALL**" title and **12 point** font for all the other annotations.

When finished **Save** your drawing as **PLASTIC REVOLVING BALL.slddrw** in your designated folder. The finished drawing is shown in **Figure 9-55** in a **Print Preview**. Then **Print** a hard copy of the drawing to submit to your instructor.

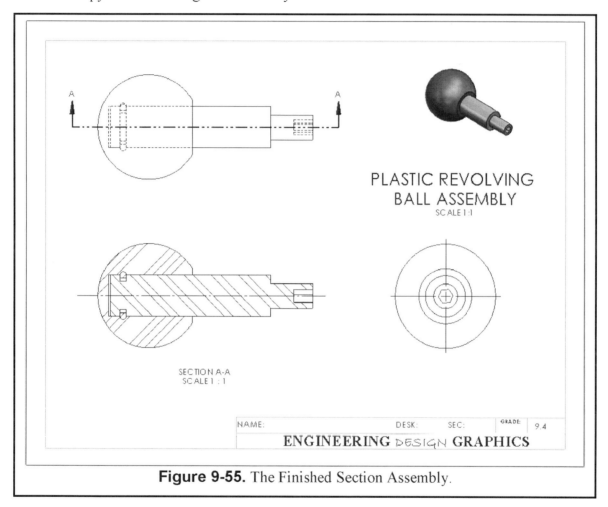

Figure 9-55. The Finished Section Assembly.

SUPPLEMENTARY EXERCISE 9-5: CLAMPING BLOCK

Build a solid model of the figure below. Make a drawing and provide a **Section** in the place of the front view. Insert a small isometric view in the upper right-hand corner of the sheet. Provide the proper Titles, Scales and other pertinent notes.

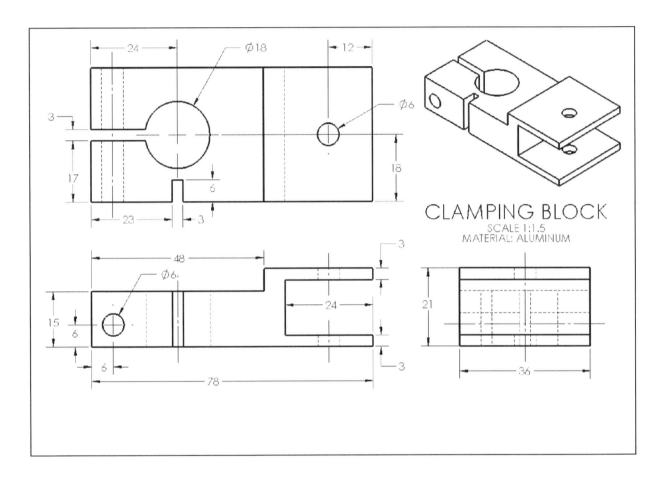

CLAMPING BLOCK
SCALE 1:1.5
MATERIAL: ALUMINUM

ALL DIMENSIONS ARE mm.

SUPPLEMENTARY EXERCISE 9-6: TWO WAY BENCH BLOCK

Build a solid model of the figure below and make a drawing. In place of the Front view add a section view of the Top view. Draw an offset line through the two counterbored holes and the center hole. Next select the menu **Insert, Drawing View, Section View** and place it in place of the Front view. From the Section view, project the Right view. Finally add an Isometric View in the upper right corner. Add notes with the Title and Scale as in previous exercises.

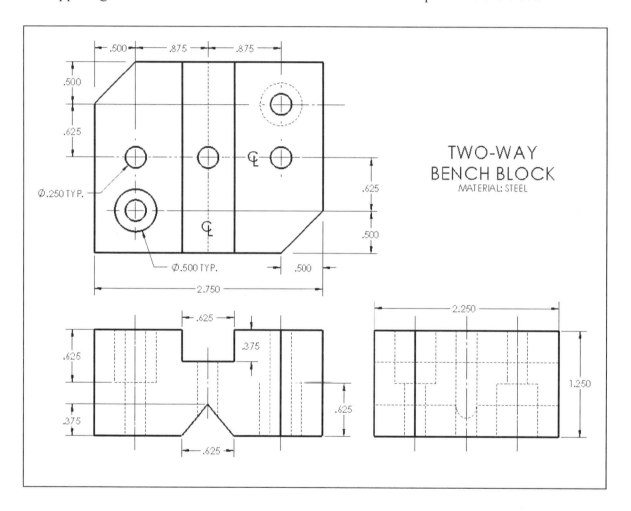

ALL DIMENSIONS ARE INCHES.

SUPPLEMENTARY EXERCISE 9-7: ELECTRICAL CONTACT PLATE

Build a solid model of the figure below and make a drawing. In place of the Front view add a section view of the Top view. Draw an offset line through the one of the small holes on the left, then through the slot with the two holes in the center and through one of the slots on the right. Next select the menu **Insert, Drawing View, Section View** and place it in place of the Front view. From the Section view, project the Right view. Finally add an Isometric View in the upper right corner. Add notes with the Title and Scale as in previous exercises.

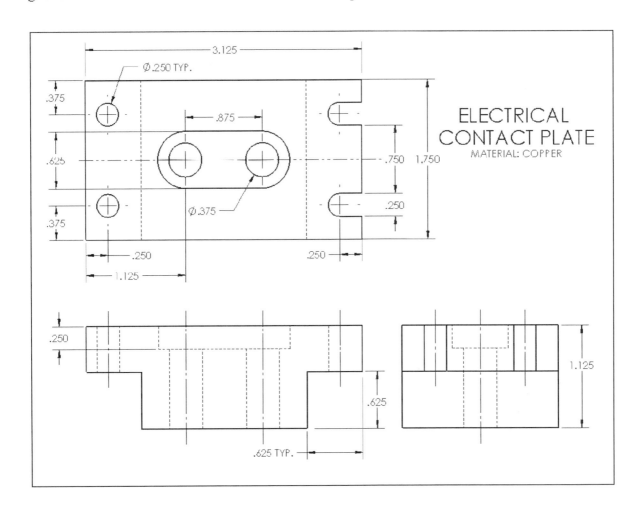

ALL DIMENSIONS ARE INCHES.

SUPPLEMENTARY EXERCISE 9-8: DISC ASSEMBLY

Build solid models of the **Disk Assembly Pieces** below. Build a fully constrained Assembly of the parts and make a drawing of it.

Provide a **Section View** in the place of the **Front** view. Finally, add an **Isometric** View in the upper right corner of the unexploded assembly in the upper right corner of your drawing. Add notes with the Title and Scale as in previous exercises.

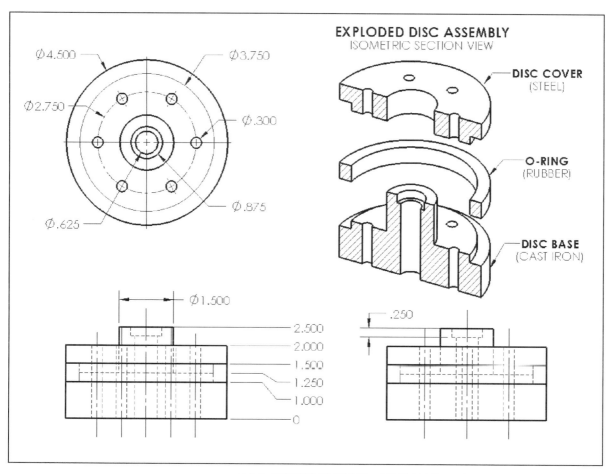

ALL DIMENSIONS ARE INCHES.

NOTES:

Computer Graphics Lab 10: Generating and Dimensioning Three-View Drawings

In Computer Graphics Lab 10, you will learn how to generate a 2D detail drawing of orthographic views from parts and assemblies. In each exercise, you will build a 3D solid model of a part, import the dimensions used in the 3D model into your 2D detail drawing, and how to manually add any missing dimensions and annotations. The following introduction will teach you the necessary tools to make a complete part or assembly drawing using SOLIDWORKS.

THE SOLIDWORKS DRAWING SHEET AND SHEET FORMAT

In chapter 1 you created a 2D drawing template called **TITLEBLOCK-INCHES.drwdot**, shown in **Figure 10-5**. In this drawing template you added a Title Block with information about the drawing such as student name, desk, section, exercise number, etc. In industrial and manufacturing drawings, a title block usually has more information including the drawing's scale, material, weight, vendor, part number, notes, revision level, etc. to correctly fabricate the part or assembly. Multiple drawing Sheets can be added to a drawing as needed.

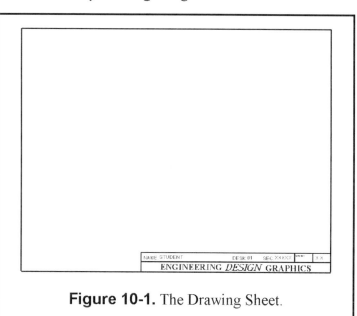

Figure 10-1. The Drawing Sheet.

The sheet format is a "locked" layer, so it cannot be accidentally modified while you make a 2D drawing, but can be easily accessed to make changes if needed. *Note:* When we modify the sheet format, the drawing views are not visible.

DETAIL DRAWING OPTIONS

Before you start adding drawing views, there are some System options that you need to set. Select the menu **Tools, Options,** select the **System Options** tab, **Drawings**, **Display Style**, where you can set the default Display Style for new detail drawing views. Select the

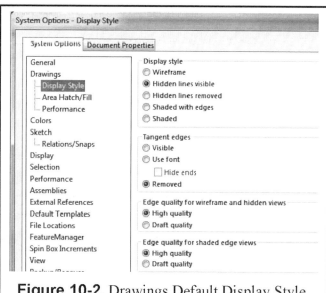

Figure 10-2. Drawings Default Display Style.

options "Hidden lines visible" and "Tangent Edges Removed" as shown in **Figure 10-2**.

The **Options** set in the **System Options** tab control the behavior and functionality of SOLIDWORKS regardless of the model open, like Sketch options, display style, file locations, etc. The options in the **Document Properties** tab are specific to the model currently open in SOLIDWORKS, including the **Drafting Standard** used (which controls annotation styles, line types and weights, arrow style and size, symbols, etc. and seldom need to be altered), units of measure, material properties, etc.

ADDING MODEL VIEWS TO A DRAWING

When you have a part or assembly open, and start a new drawing, activate the drawing sheet, select the **View Palette** tab where you can see a preview of each orientation of the model open in **SOLIDWORKS**, or you can select a different model from the drop-down list at the top of the **View Palette**.

It is advisable to turn on the **Auto-Start Projected View** option at the top as shown in **Figure 10-3**. From here we can drag any view orientation needed, and automatically project the other dimensions from it. If this option is not activated, you will need to add the different views individually, or use the **Projected View** command from the **View Layout** tab after adding one view.

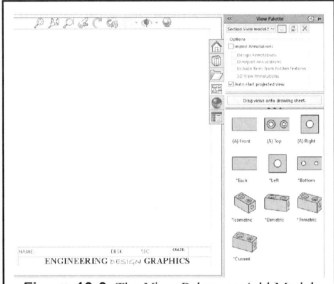

Figure 10-3. The View Palette to Add Model Views to a Drawing.

From the **View Palette** drag the **Front** view onto the drawing, and move the cursor around to see the other projected views. Click above to add a **Top** view, on the upper right to add an **Isometric** view, and to the right side to add a **Right** view, as shown in **Figure 10-4**.

The views will be created with the Hidden Lines Visible system option you set before.

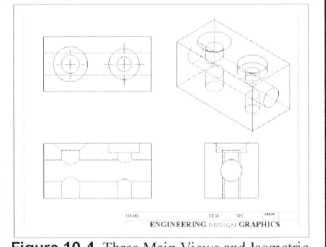

Figure 10-4. Three Main Views and Isometric.

If the drawing views are too big for the sheet, right click on *Sheet1* in the FeatureManager, or in the drawing (*Not a model view*) and select **Properties** to change the scale of the entire sheet.

Select the **Isometric** view. In the view's properties listed in the PropertyManager change the view's scale to make it smaller, and change the **Display Style** to **Shaded with Edges** mode. Now you can click and drag to move the views around the drawing sheet to have enough space for dimensions and annotations. The orthographic views will remain aligned as you move them.

DIMENSIONING THE DRAWING

After adding the necessary views to your drawing, you need to add dimensions and annotations to the different views. When you make a 3D part, sketch and feature dimensions are added. These are called **Parametric** dimensions and can be imported into a detail drawing, making it easier to fully annotate the drawing. When parametric dimensions are changed in the part *or* the drawing, both the 3D model and the 2D drawing views are updated accordingly. To **import** the 3D model dimensions and annotations into a 2D drawing, from the **Annotation** tab select the **Model Items** command.

Figure 10-5. Model Items Options.

In the **Model Items** command options, you can select to import into a selected view or all drawing views; from the entire 3D model or a selected feature, we can select the type of dimensions, annotations and reference geometry to import from the 3D model, as shown in **Figure 10-5**. Some of the annotations available to import into a drawing include:

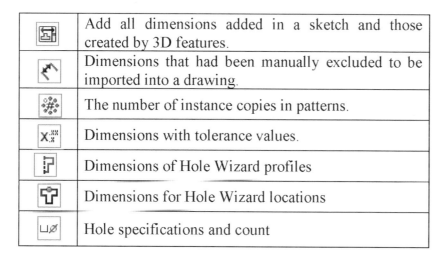	Add all dimensions added in a sketch and those created by 3D features.
	Dimensions that had been manually excluded to be imported into a drawing.
	The number of instance copies in patterns.
	Dimensions with tolerance values.
	Dimensions of Hole Wizard profiles
	Dimensions for Hole Wizard locations
	Hole specifications and count

If the imported 3D model dimensions do not fully describe the detail drawing, you can manually add the missing necessary dimensions. By default, manually added dimensions have a parenthesis, which is interpreted as a reference dimension, meaning its value is not exact. A dimension's parentheses can be turned on or off individually in the dimension's properties. To turn off the parentheses for all manually added dimensions, you need to go to the menu **Tools, Options, Document Properties**, select the **Dimensions** section and turn off the "**Add Parentheses by default**" option. Remember this is a document option, not a system option. You would have to set this option for each drawing individually, or to the template used to make these drawings.

After you import the model dimensions and hole callout annotations, the dimensions can be arranged in the different views for visibility. To **move a dimension** from one view to another that would better describe the model, you have to drag the dimension to the other view while holding the **Shift** key.

After you add the **Center Mark** and **Centerlines** to the drawing, and arranging the dimensions, the drawing looks like **Figure 10-6**.

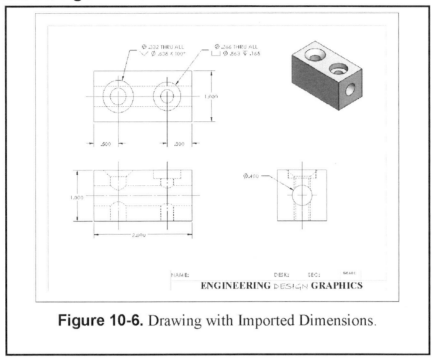

Figure 10-6. Drawing with Imported Dimensions.

Now you can manually add the vertical location dimensions of the holes using the **Smart Dimension** tool, and turn off the parentheses. Using the **Note** command from the **Annotations** tab add the **Title**, **Scale** and **Material** for the part. The finished drawing looks like **Figure 10-7**.

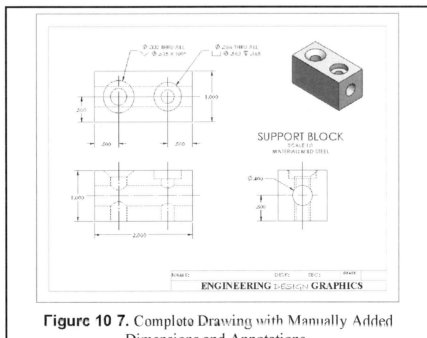

Figure 10 7. Complete Drawing with Manually Added Dimensions and Annotations.

Exercise 10.1: GUIDE BLOCK Drawing

In Exercise 10.1, you will make a solid model of the Guide Block using SOLIDWORKS commands that you have learned in previous labs. Next, you will make a drawing with three orthographic views and an Isometric of the part. You will import the 3D model dimensions and complete the drawing with annotations and a title.

BUILDING THE GUIDE BLOCK PART

Open **ANSI-INCHES.prtdot** and immediately **SAVE AS – GUIDE BLOCK.sldprt**. Select the **TOP PLANE** and change to a **Top** view orientation. Use **Inch** "Units" to **3** decimal places. Add a new **Sketch** in the Top plane, draw and dimension the profile shown in **Figure 10-8**. Draw a construction line and use the Dynamic Mirror tool to make the sketch symmetric. Then **Extrude** the profile *upward* from the plane using a **Blind** end condition of **.75** inches.

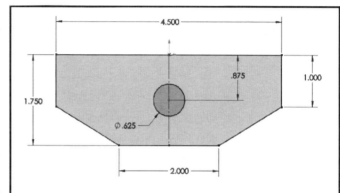

Figure 10-8. The Sketch for the Base Feature of the Guide Block.

Change to a **Back** view orientation, select the back face and add a new sketch in it. Draw a **Rectangle** and **Dimension** it as shown in **Figure 10-9**. Add a collinear relation between the edge of the base and the rectangle. Draw a vertical **Centerline** through the origin and **Mirror** the rectangle to the other side. Change to an **Isometric** view, and **Extrude** the sketch *forward* using a **Blind** end condition of **1.00** inches.

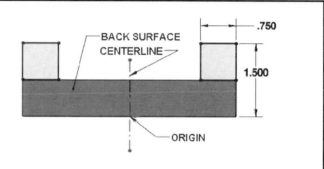

Figure 10-9. Sketching on the Back Surface.

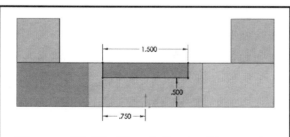

Figure 10-10. Sketch for the Slot on the Front Face.

The final feature for the Guide Block is a guide slot that goes all the way through from front to back. Change to a **Front** view orientation, select the front face of the part and add a new **Sketch.** Add a **Rectangle** and **Dimension** it as shown in **Figure 10-10**. Then use **Extruded Cut** to cut the sketch using a **Through All** condition in the *backward* direction. When finished, right click on **Material** in the FeatureManager and assign **AISI 1020** from the **Steel** library.

The Guide Block solid model is now complete, as shown in **Figure 10-11**. Now **Save** your part as **GUIDE BLOCK.sldprt** in your designated folder, but *do not close* your file. You will now make the detail drawing.

MAKE A MULTI-VIEW DRAWING

From your folder, open the **TITLEBLOCK-INCHES.drwdot** Drawing template. Immediately **Save as GUIDE BLOCK.slddrw**.

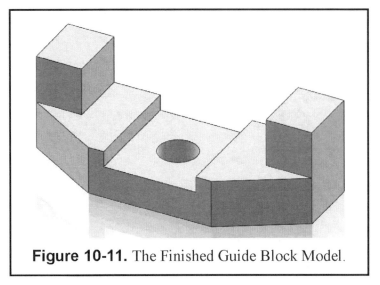

Figure 10-11. The Finished Guide Block Model.

Open the **View Palette**, make sure the **Auto-Start Projected view** option is turned on and the **Guide Block** is selected in the drop-down list.

Select the **Front** view and drag it into the drawing sheet, move up to add the **Top** view, to the top-right to add an **Isometric** view, and to the right to add a **Right** view. If needed, right click in **Sheet1** on the FeatureManager, select **Properties** and change the scale to 1:1.

Select the **Isometric** view and change its scale to 1:2, and set it to **Shaded with Edges** display mode, and add the Centerlines to continue. Your drawing now looks like **Figure 10-12**.

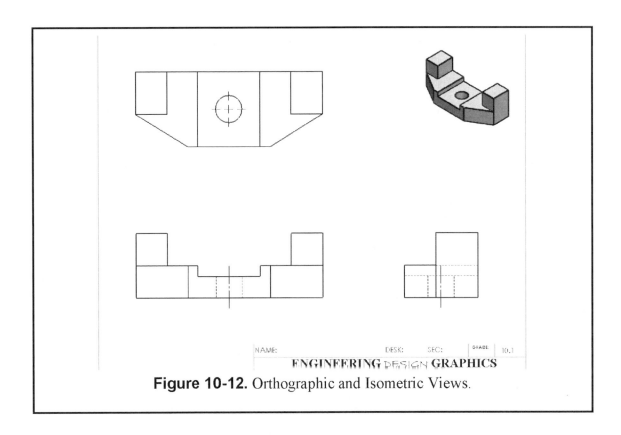

Figure 10-12. Orthographic and Isometric Views.

DIMENSIONING THE DRAWING

From the **Annotations** tab, select the **Model Items** command. In the Source section select **Entire Model,** turn on the option **Import items into all views** and click **OK**. Your 2D drawing with the imported 3D model dimensions should look like **Figure 10-13.**

To better distribute the dimensions, select the **0.750in** vertical dimension in the **Front** view, hold down the **Shift** key and drag it to the **Right** view. In the **Top** view, there are two **1.000in** dimensions, and one of them is redundant. Select one and delete it.

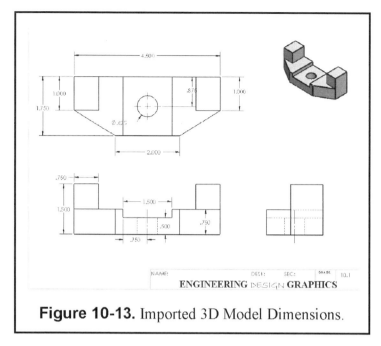

Figure 10-13. Imported 3D Model Dimensions.

Since the part was made symmetric about the origin, some features are not dimensioned. We can manually add those dimensions. Select the **Smart Dimension** tool, and add the missing dimensions to match the completed drawing in **Figure 10-14**. Remember to turn off the parentheses to the manually added dimensions in the dimension's properties. Finally add a **Note** with the Title, Scale and Material. **Save** your drawing as **GUIDE BLOCK** and **Print** a hard copy to submit to your instructor.

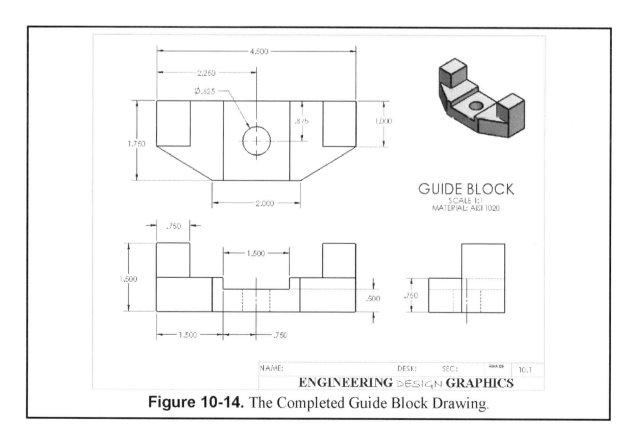

Figure 10-14. The Completed Guide Block Drawing.

Exercise 10.2: PIPE JOINT Drawing

In this Exercise 10.2, you will build a solid model of the Pipe Joint using SOLIDWORKS commands that you have learned in previous labs. This exercise will be made using millimeter units. Next, you will make a drawing with three orthographic views and an Isometric of the part. You will import the 3D model dimensions and complete the drawing with annotations and a title.

BUILDING THE PIPE JOINT MODEL

Open the **ANSI-METRIC.prtdot** and **Save As PIPE JOINT.sldprt**. Select the **Top** plane, add a **Sketch** and change to a **Top** view orientation. In the sketch use the **Circle** tool to draw two circles as shown in **Figure 10-15**. The first circle is centered at the origin; the other circle is to the right. **Dimension** the diameters as shown (φ**56** mm and φ**28** mm). **Dimension** the two circles **48** mm apart. Now **Add** a **Horizontal Relation** between the centers of the two circles so that they are aligned. Next, draw two **Lines** that are approximately tangent to both circles on top and bottom as shown in **Figure 10-15**. Then **Add** four **Tangent Relation** between the lines and circles as shown.

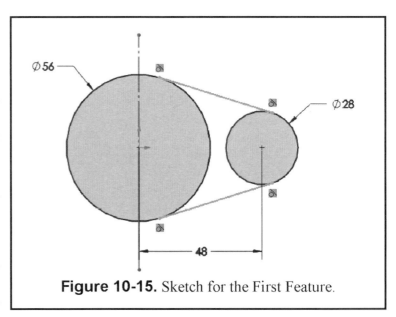

Figure 10-15. Sketch for the First Feature.

Finally add a vertical **Centerline** that goes through the origin and extends beyond the big circle's perimeter, as in **Figure 10-16**. Select the **Trim Entities** tool and remove the left side of the small circle, the left side of the large circle where it intersects the centerline, and the right side of the big circle where

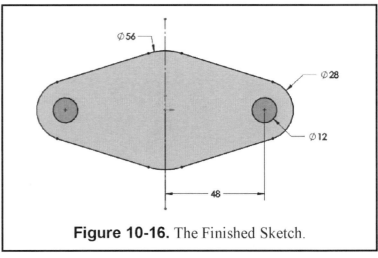

Figure 10-16. The Finished Sketch.

it is tangent to the top and bottom lines. **Draw** a circle concentric to the small arc with a **Diameter of 12 millimeters**. Select the **Mirror** command to copy the entire sketch about the vertical centerline to finish, as shown in **Figure 10-16**.

Now **Extrude** the Sketch up using a **Blind** end condition of **16** mm to complete the base feature for the Pipe Joint, as shown in **Figure 10-17**.

Select the top face, add a **Sketch** and draw two **Circles** concentric to the holes in the base. Add a relation to make them **Equal**, and dimension one **18** mm in diameter. Then make an **Extruded Cut 6** mm into the base feature to make a counterbored hole on each end of the base plate, as shown in **Figure 10-17**.

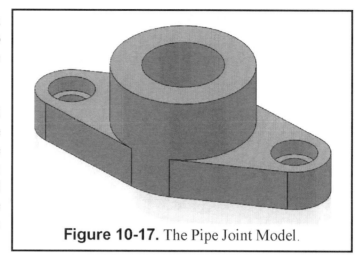

Figure 10-17. The Pipe Joint Model.

In a **Top** view, add a new sketch in the same face, and draw a **Circle** centered in the origin. Add an **Equal** relation with the **56** mm round edge. Now add an **Extruded Boss** with a **Blind** distance of **24** mm.
Now select the top face of the new Boss, add a **Sketch** and draw a **Circle** centered in the origin with a **32** mm diameter. Make an **Extruded Cut** using a **Through All** end condition. Your part should now look like **Figure 10.17**.

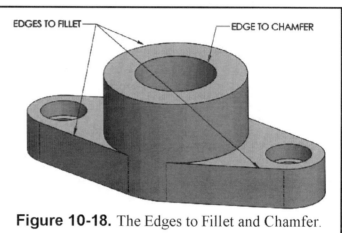

Figure 10-18. The Edges to Fillet and Chamfer.

The final step is to round several model edges. Select the **Fillet** command, set the radius to **3** mm and select the rounded edges shown in **Figure 10-18** and click **OK**. Now add a 3 mm chamfer to the inside edge at the top of the hole. To finish, set the material to **Wrought Stainless Steel**.

The Pipe Joint part is now finished, as shown in **Figure 10-19**. **Save** your part as **PIPE JOINT.sldprt** in your designated folder, but do not Close your file. You will now make the detail drawing.

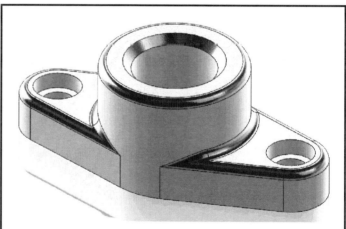

Figure 10-19. The Finished Pipe Joint Model with Fillets.

MAKE A MULTI-VIEW DRAWING

From your folder, open the **TITLEBLOCK-METRIC.drwdot** drawing template. Immediately select **Save AS** to save the new drawing as **PIPE JOINT.slddrw**.

Before you start the drawing, select the menu **Tools, Options**. In the **Document Properties, Detailing** section set the Dimensioning Standard to **ANSI,** and click **OK** to finish.

Open the **View Palette**; for this lab, make sure the **Auto-Start Projected view** option is turned **off** and the **Pipe Joint** is selected in the drop-down list.

From the **View Palette** drag a **Top** view, and then an **Isometric** view. Select the **Isometric** view and change it to **Shaded with Edges** mode. If needed, right click on the **Top** view and select **Tangent Edge, Tangent Edges Removed**, and change the Top view's scale to 1:1, as shown in **Figure 10-20.**

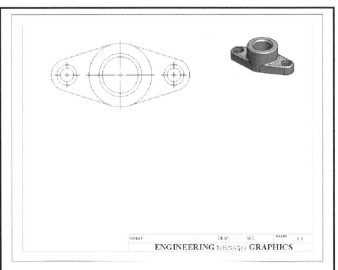

Figure 10-20. Top and Isometric Views of the Pipe Joint.

From the **View Layout** tab select the **Section View** command. Activate the **Horizontal** section view option, and click in the **Top** view when the section line snaps in the middle of the part. Immediately locate the section view below in place of the **Front** view.

Now select the **Section** view, from the **View Layout** tab select the **Projected View** command, and locate the projected **Right** on the right side of the sheet. Select the **Right** view, and set the **Display Style** to **Hidden Lines Visible**. Add the **Centerlines** to continue. The final drawing layout is shown in **Figure 10-21**.

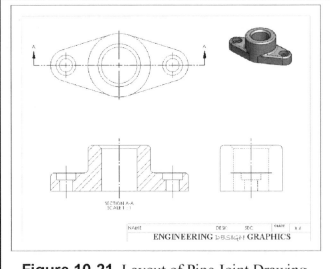

Figure 10-21. Layout of Pipe Joint Drawing.

DIMENSIONING THE DRAWING

From the **Annotations** tab, select the **Model Items** command. In the Source section select **Entire Model,** turn on the option **Import items into all views** and click **OK**. When you import dimensions, they are first added to the **Detail** views, then **Section** views, and finally to main views. Move some of the dimensions from the section view to the **Top** and **Right** views. Remember to hold down the **Shift key** to move dimensions to a different view. Delete the dimensions of the counterbore hole (both diameters and the depth). Your 2D drawing should look like **Figure 10-22.**

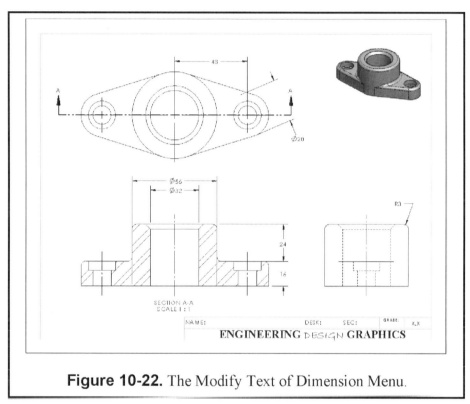

Figure 10-22. The Modify Text of Dimension Menu.

After moving the 28mm Diameter dimension it will be shown as a linear dimension. To change it to a radial dimension right click on it, and select **Display Options, Display as Diameter**. To show it as a **Radius** dimension, right click on it again and select **Display Options, Display as Radius**.

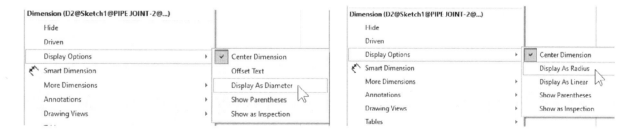

Since this dimension references both sides of the part, select the **R14.00 mm.** In the dimension's properties add a "**2X**" before "**R<DIM>**" in the **Dimension Text box**. The final dimension should read "**2X R14**".

With the **Smart Dimension** tool, add a dimension between the two counterbore holes. From the Drop-down **Smart Dimension** command, select **Chamfer Dimension**. In the **Section** view select the chamfer edge, then the vertical edge of the hole and locate the Chamfer dimension.

Finally, from the **Annotation** tab select the **Hole Callout** command and click on the outer edge of the counterbore hole in the **Top** view. Locate the annotation to finish. As with the radial dimension, add a "**2X**" in the **Dimension Text box**.

Next use the **Note** command to add a note with title, scale, material, "**ALL FILLETS & ROUNDS -R3**", and "**UNITS MM**", as shown in **Figure 10-23.** Change the font and size as needed.

Save your drawing and name it **PIPE JOINT.slddrw**, and **Print** a hard copy to submit to your lab instructor.

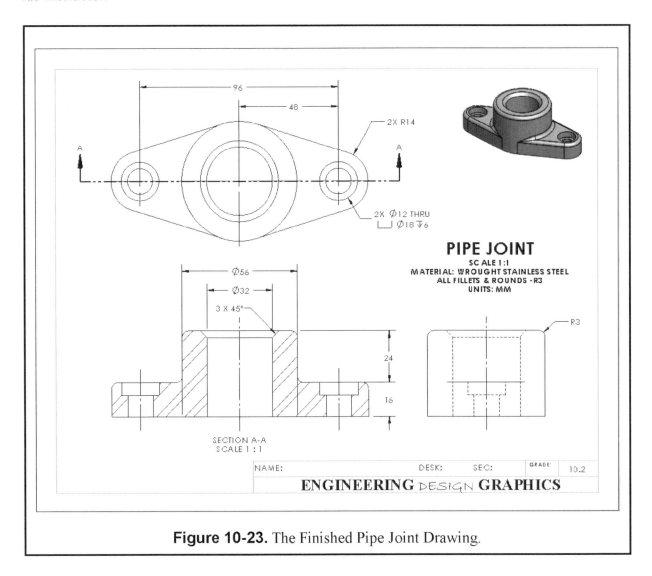

Figure 10-23. The Finished Pipe Joint Drawing.

Exercise 10.3: PEDESTAL BASE Drawing

Quite often in mechanical design, features such as holes are repeated in a regular circular pattern. Examples include flanged pipe joints, inspection cover plates, portal bezels, motor end housings, and many others. Circular and rectangular patterns have been introduced in earlier exercises; here, we create a part with multiple bolt circle patterns and discuss dimensioning practices for the part in a multi-view drawing. You will create a Pedestal Base typical of a stage in a precision manufacturing tool.

BUILDING THE PEDESTAL BASE

Start a new part using the METRIC template, and save it as **PEDESTAL BASE.sldprt**.

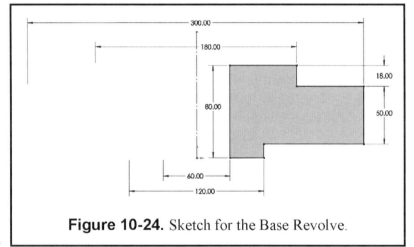

Figure 10-24. Sketch for the Base Revolve.

Select the **Front Plane** and add a new **Sketch**. Change to a **Front** view and draw the sketch profile shown in **Figure 10-24**. This sketch will be used to make a **Revolved Boss**, to add diameter dimensions to the sketch select the centerline, the line to dimension, and finally cross the centerline **BEFORE** locating the dimension; it will be automatically doubled. While you add the **doubled dimensions**, the mouse cursor will be shown as: When you finish adding the doubled dimensions press the **ESC** key to return to the **Smart Dimension** tool to finish adding the height dimensions.

When the sketch is complete, add a **360** deg **Revolved Base** and click **OK**. The resulting part will look like **Figure 10-25**.

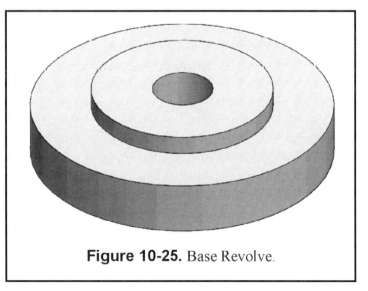

Figure 10-25. Base Revolve.

Next, create the circular hole patterns in the pedestal base. The holes needed can be made using the **Hole Wizard**, which will be covered in the next lab, and for this exercise we'll use a regular **Cut Extrude** feature.

Change to a **Top** view and add a new **Sketch** in the outer top face of the base feature. Draw two construction lines and a construction circle. After adding the 240mm circle select it, and in the PropertyManager select the checkbox **For Construction**. Finally dimension it as shown in **Figure 10-26**.

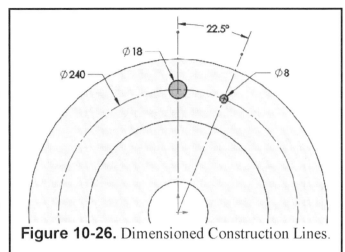

Figure 10-26. Dimensioned Construction Lines.

When the sketch is complete select the **Cut Extrude** command and use the **Through All** end condition.

From the **Features** tab, select the **Circular Pattern** command, select the circular outer edge for the direction, add the **Cut Extrude** feature in the **Features to pattern** selection box, set the number of copies to **8** and use the equal spacing option in **360 degrees**.

To finish the part, add a **5mm Fillet** to the outer edges, except for the holes, as shown in **Figure 10-27**.

Edit the part's material and set it to **Copper** from the Copper Alloys library. Save your part **File** as **PEDESTAL BASE.sldprt**.

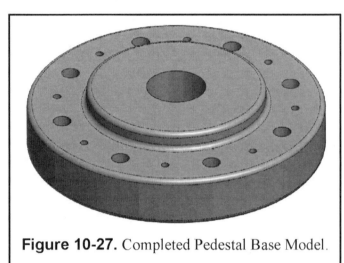

Figure 10-27. Completed Pedestal Base Model.

MAKE A MULTI-VIEW DRAWING

Using the **METRIC** drawing template start a new drawing and save it as **Pedestal Base.slddrw**.

Open the **View Palette**, for this lab, make sure the **Auto-Start Projected view** option is turned **off** and the **Pedestal Base** is selected in the drop-down list.

From the **View Palette** drag a **Top** view, and then an **Isometric** view. Select the **Isometric** view and change it to **Shaded with Edges** mode. If needed, right click on the **Top** view and select **Tangent Edge, Tangent Edges Removed**, and change the Top view's scale to 1:4.

From the **View Layout** tab select the **Section View** command. Activate the **Horizontal** section view option, and click in the **Top** view when the section line snaps in the middle of the part. Immediately locate the section view below in place of the **Front** view.

Now select the **Section** view, from the **View Layout** tab select the **Projected View** command, and locate the projected **Right** on the right side of the sheet. Select the **Right** view, and set the **Display Style** to **Hidden Lines Visible**. Add the **Centerlines** to continue.

Now you need to import the part's dimensions into the drawing. From the **Annotations** tab, select the **Model Items** command. In the Source section select **Entire Model,** turn on the option **Import items into all views** and click **OK**.

When you import dimensions, they are first added to the **Detail** views, then **Section** views, and finally to main views.

Move some of the dimensions from the section view to the **Top** and **Right** views. Remember to hold down the **Shift key** to move dimensions to a different view. **Delete** the diameter dimensions of the two holes. Your 2D drawing should look like **Figure 10-28.**

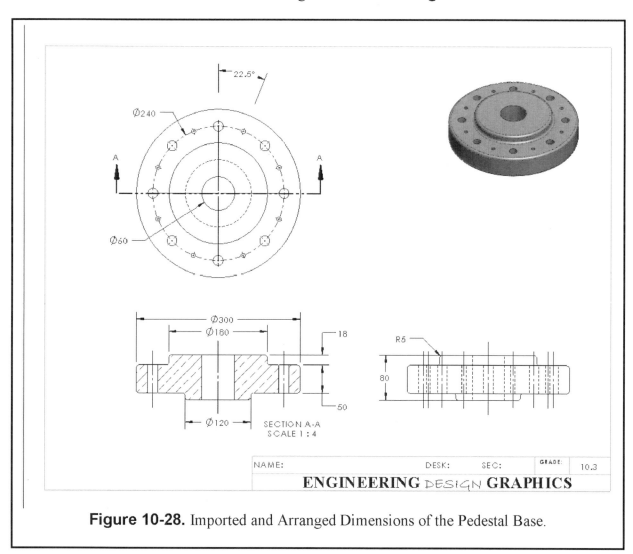

Figure 10-28. Imported and Arranged Dimensions of the Pedestal Base.

Now, add a **centerline** in the **Top** view from the origin to the first small hole to reference the imported **22.5 deg** dimension.

The **Fillet** dimension is added to a single fillet. To communicate the fillet's dimension is the same for all rounded edges, select the dimension, and in the **Dimension Text** box type "**TYP.**" at the end of the dimension text.

To complete the drawing, select the **Hole Callout** tool and click in both of the patterned holes; the correct annotation will be added. In the **Dimension Text** of both hole annotations type "**8X**" before the dimension text code.

Next use the **Note** command to add a note with title, scale, material, and "**UNITS MM**", as shown in **Figure 10-29.** Change the font and size as needed.

Save your drawing and name it **PEDESTAL BASE.slddrw**, and **Print** a hard copy to submit to your lab instructor.

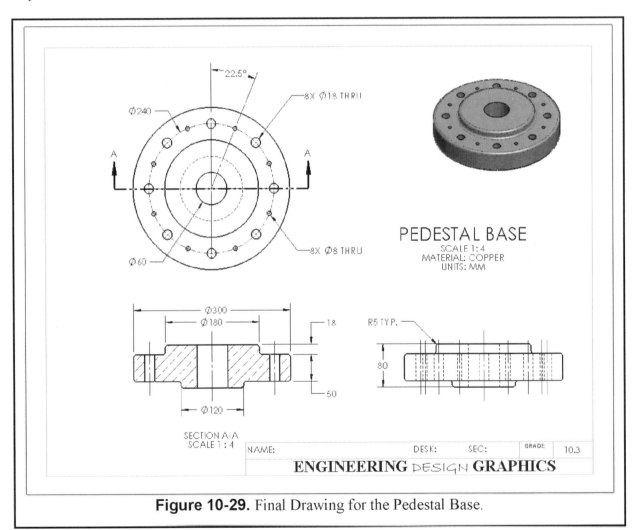

Figure 10-29. Final Drawing for the Pedestal Base.

Exercise 10.4: TOOLING PAD Drawing

In Exercise 10.4, you will build a solid model of the Tooling Pad using SOLIDWORKS commands that you have learned in previous labs. Next, you will make a drawing with three orthographic views of the part. You will import the model dimensions and add the necessary annotations to complete the drawing.

BUILDING THE TOOLING PAD MODEL

Start a new part using the INCH template, and save it as **TOOLING PAD.sldprt**.

Add a new sketch in the **Top** plane, draw the profile in **Figure 10-30** using the **Center Rectangle** tool from the origin, and dimension as shown. Then **Extrude** the profile up from the plane using a **Blind** end condition of **0.725** inches.

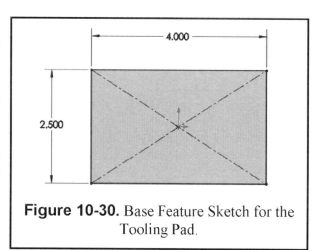

Figure 10-30. Base Feature Sketch for the Tooling Pad.

Adding a Hole

Figure 10-31. The Hole Wizard.

Now you will use the Hole Wizard to add some holes on the Tooling Pad. From the Features tab select the **Hole Wizard** command. In the "**Hole Specification**" select the Hole option, and set the following parameters, as indicated in **Figure 10-31**:

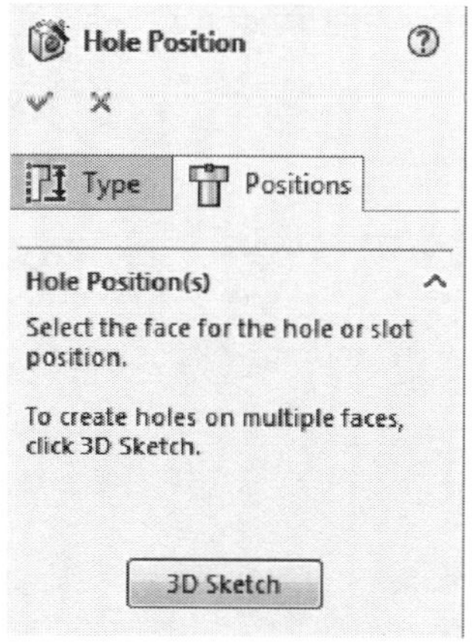

Standard: **Ansi Inch**
Type: **Fractional Drill Sizes**
End Condition: **Through All**
Hole Diameter: **3/8 in**

After the hole parameters have been set, select the **POSITIONS** tab; this is the second part of the Hole Wizard as shown in **Figure 10-32**. Change to a **Top** view and select the top face of the base feature. In this step of the **Hole Wizard** we are editing a sketch, and you have to add a sketch **Point** to define the location of each hole.

Figure 10-32. Hole Position.

After you select the top face, the sketch **Point** tool is automatically selected. Add three points in the top face and dimension them as shown in **Figure 10-33**. At each **Point** location you will see a preview of the hole. When you finish dimensioning the locations click **OK** to finish.

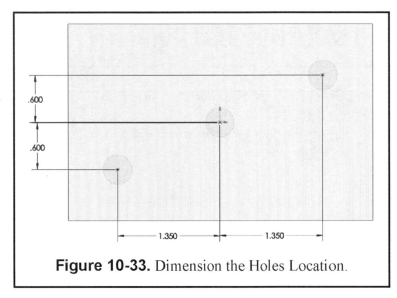

Figure 10-33. Dimension the Holes Location.

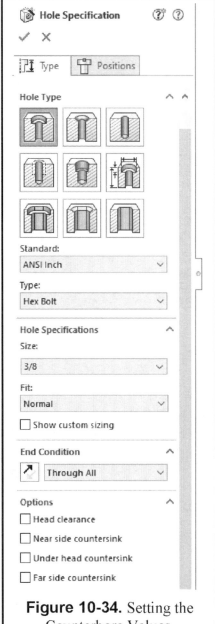

Figure 10-34. Setting the Counterbore Values.

Adding a Counterbore

Now you will add a counterbore hole using the **Hole Wizard**. In the **Hole Specification** select the **Counterbore** option, and set the following parameters, as indicated in **Figure 10-34.**

> Standard = **Ansi Inch**
> Type = **Hex Bolt**
> Size = **3/8 in**
> End Condition = **Through All**

Selecting **Show Custom Sizing** will show the defaults for this size bolt. Make sure that all the boxes under **Options** are unselected.

Now select the **Positions** tab. This brings you to the second part of the Hole Wizard as shown in **Figure 10-35**. Change to a **Top** view and select the top face of the base feature.

Add a sketch **Point** to define the location of each counterbore hole.

After you select the top face, the sketch **Point** tool is automatically selected. Add two points in the top face and dimension them as shown in **Figure 10-35**. At each **Point** location you will see a preview of the hole. When you finish dimensioning the locations click **OK** to finish.

When the **Hole Wizard** is complete, add a **Chamfer** of **0.02** inches in the top edges of the counterbore holes, as seen in **Figure 10-36.**

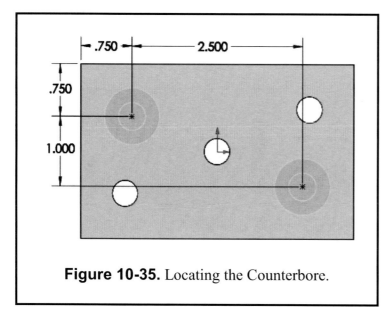

Figure 10-35. Locating the Counterbore.

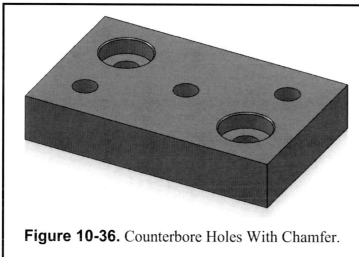

Figure 10-36. Counterbore Holes With Chamfer.

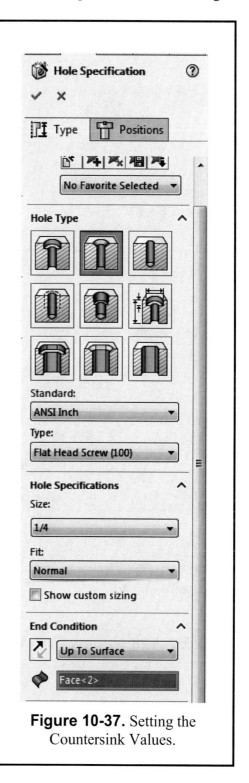

Figure 10-37. Setting the Countersink Values.

Adding a Countersink

Now you will add a countersink hole using the **Hole Wizard.** In the **Hole Specification** select the **Countersink** option, and set the following parameters, as indicated in **Figure 10-37.**

Standard = **Ansi Inch**
Type = **Flat Head Screw**
Size = **1/4 in**
End Condition = **Up to Surface**

*For the Up to Surface end condition select the inner face of the small middle through hole.

Now select the **Positions** tab. This brings you to the second part of the Hole Wizard. Change to a **Front** view and select the front face of the base feature.

Click to add a Point in the Front face and dimension it as shown in **Figure 10-38**. Click OK to finish the Hole Wizard.

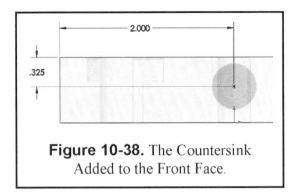

Figure 10-38. The Countersink Added to the Front Face.

Adding the Bottom Base Feet

Change to a **Bottom** view to add the four corner feet to the Tooling Pad. Add a **Sketch** in the bottom face and draw the profile as shown in **Figure 10-39**.

Draw four **Rectangles** in the four corners, select all the lines in the sketch and add an Equal relation to make them the same, and then add a **.325** dimension to any of the lines. Next, **Extrude** the sketch **.125** inches away from the model. Add **.0625 Fillets** to the outer top edges and the outer vertical edges of the tooling pad.

Edit the part's material and set it to **Titanium Ti-8Mn, Annealed** from the **Titanium Alloys** library. Save your part as **TOOLING PAD.sldprt**.

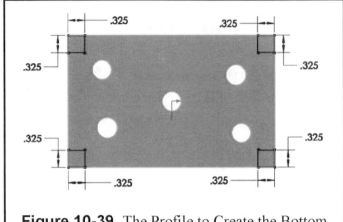

Figure 10-39. The Profile to Create the Bottom Feet of the Tooling Pad.

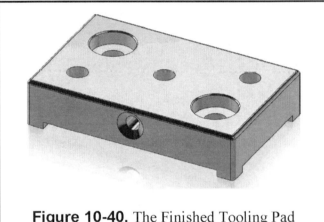

Figure 10-40. The Finished Tooling Pad Model.

MAKE A MULTI-VIEW DRAWING

Use the INCH drawing template start a new drawing, and **SAVE AS TOOLING PAD.slddrw,** and change the sheet's scale to 1:1.

Open the **View Palette**, for this lab, make sure the **Auto-Start Projected view** option is turned **on** and the **Tooling Pad** is selected in the drop-down list.

From the **View Palette** drag a **Front** view, then project a **Top** and an **Isometric** view. Select the **Isometric** view and change it to **Shaded with Edges** mode and change the scale to 1:2. If needed right click on the **Front** and **Top** view and select **Tangent Edge, Tangent Edges Removed**. See **Figure 10-41**.

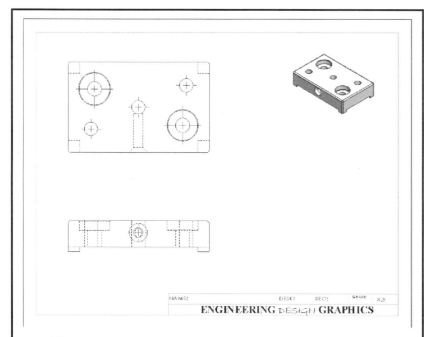

Figure 10-41. Front, Top and Isometric views of the Tooling Pad

ADD A SECTION VIEW

From the **View Layout** tab select the **Section View** command. Activate the **Vertical** section view option, and click in the **Front** view when the section line snaps in the middle of the part. Immediately locate the section view to the right of the **Front** view, as shown in **Figure 10-42**.

ADDING CENTERLINES AND CENTER MARKS

From the Annotations tab, select the **Centerline** command and add the missing centerlines to the Top, Front, and Section views. If missing, also add the missing **Center Marks**.

DIMENSIONING THE DRAWING

Now you need to import the part's dimensions into the drawing. From the **Annotations** tab, select the **Model Items** command. In the Source section select **Entire Model,** turn on the option **Import items into all views** and click **OK**.

By activating the **Hole Callout** option in the **Model Items** command, the correct parametric annotations are added to the **Hole Wizard** features, including the correct number or instances for each one. In this case, the dimensions added fully annotate the drawing. There are only a few changes to make, such as moving dimensions to a different view (hold **Shift** to move dimensions). In the bottom feet, you made all lines equal and only dimensioned one; in this case add a **TYP** text after the **.325** dimension. Keep in mind that a dimension's arrows can be **reversed** by selecting the dimension, and click in the arrow to reverse it, as shown in **Figure 10-43**. Add the missing dimensions as needed to finish.

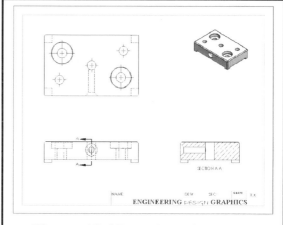

Figure 10-42. Section View of the Tooling Pad.

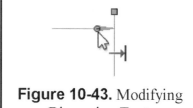

Figure 10-43. Modifying Dimension Text.

Next use the **Note** command to add a note with title, scale, material, and "**ALL FILLETS AND ROUNDS R- .0625**"", as shown in **Figure 10-44.** Change the font and size as needed.

Save your drawing and name it **TOOLING PAD.slddrw**, and **Print** a hard copy to submit to your lab instructor.

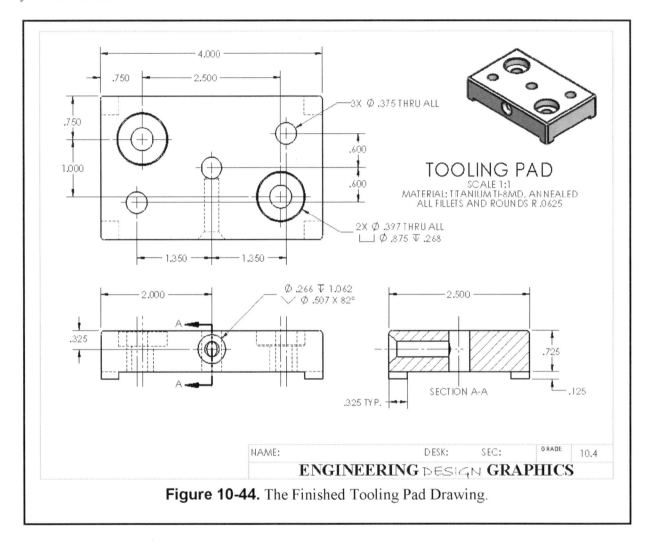

Figure 10-44. The Finished Tooling Pad Drawing.

SUPPLEMENTARY EXERCISE 10-5: PILLOW BLOCK

Build a solid model of the figure below. Make a detail drawing and Dimension it. Insert a small Isometric of the part in the upper right hand corner of the sheet. Provide the proper Titles, Scales and other pertinent notes.

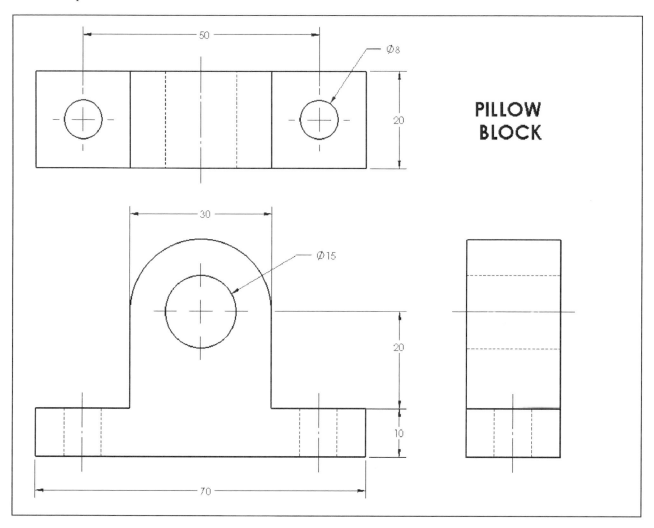

ALL DIMENSIONS ARE mm.

SUPPLEMENTARY EXERCISE 10-6: CHASSIS BOX

Build a solid model of the figure below. Make a detail drawing and Dimension it. Insert a small Isometric of the part in the upper right hand corner of the sheet. Provide the proper Titles, Scales and other pertinent notes.

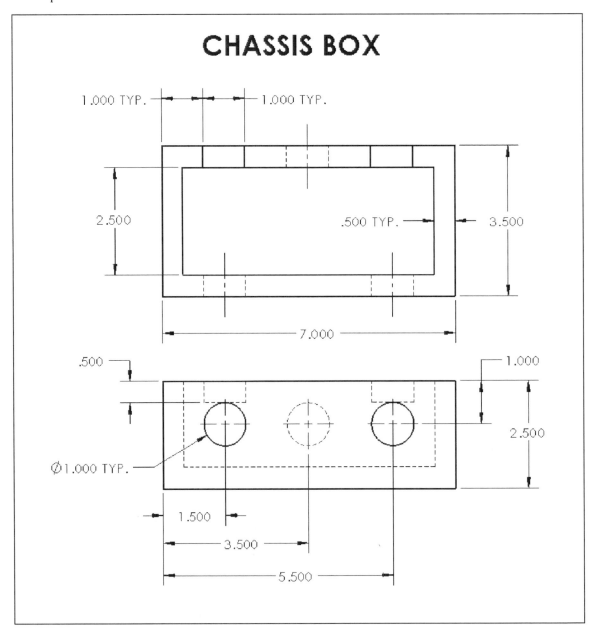

CHASSIS BOX

UNITS IN INCHES.

SUPPLEMENTARY EXERCISE 10-7: COLUMN BASE

Build a solid model of the figure below. Make a detail drawing and Dimension it. Instead of a Front view, add a Horizontal section view of the Top view. Add an Isometric of the part in the upper right hand corner of the sheet. Provide the proper Titles, Scale and other pertinent notes.

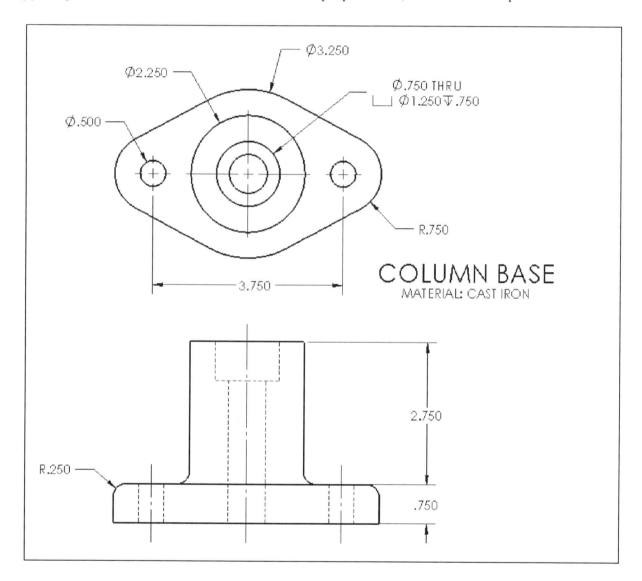

UNITS IN INCHES.

SUPPLEMENTARY EXERCISE 10-8: VALVE HOUSING

Build a solid model of the figure below. Make a detail drawing and Dimension it. Add a Section view in place of the Front view, and project the Right view from the Section view. Add a small Isometric of the part in the upper right hand corner of the sheet. Provide the proper Titles, Scales and other pertinent notes.

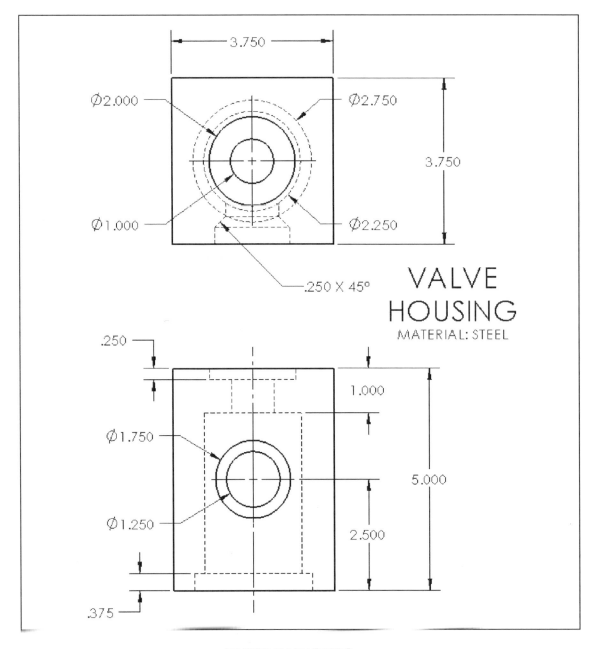

UNITS IN INCHES.

APPENDIX A
EXAMPLE OF A TILEBLOCK WITH DIMENSIONS

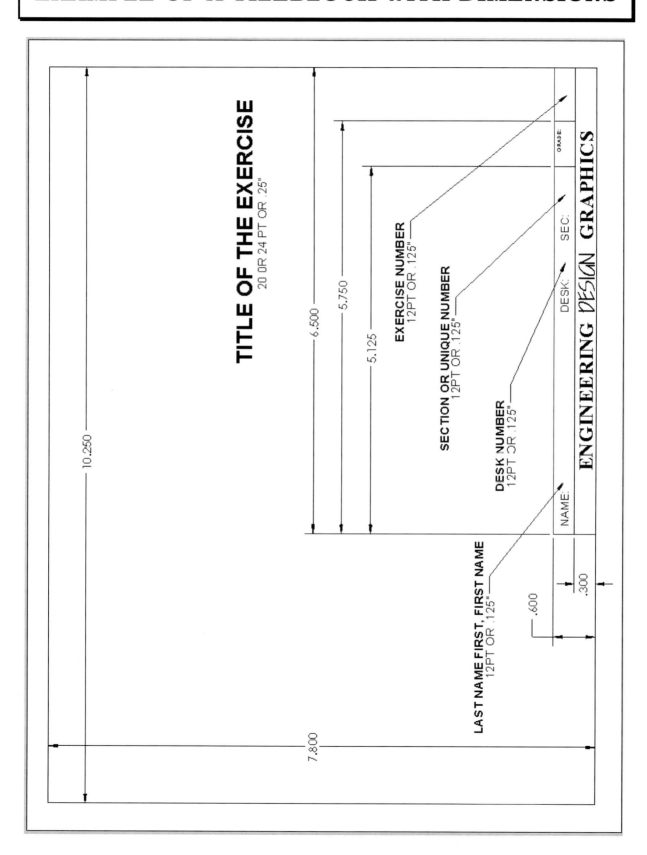

NOTES: